Technology-Driven Innovation in Gulf Cooperation Council (GCC) Countries:

Emerging Research and Opportunities

Anna Visvizi
Deree - The American College of Greece, Greece & Effat University, Saudi Arabia

Saad Bakry
King Saud University, Saudi Arabia

Miltiadis D. Lytras
Deree - The American College of Greece, Greece & Effat University, Saudi Arabia

Wadee Alhalabi
Effat University, Saudi Arabia

A volume in the Practice, Progress, and Proficiency in Sustainability (PPPS) Book Series

IGI Global
DISSEMINATOR OF KNOWLEDGE

Published in the United States of America by
IGI Global
Business Science Reference (an imprint of IGI Global)
701 E. Chocolate Avenue
Hershey PA, USA 17033
Tel: 717-533-8845
Fax: 717-533-8661
E-mail: cust@igi-global.com
Web site: http://www.igi-global.com

Library of Congress Cataloging-in-Publication Data

Names: Visvizi, Anna, 1976- editor.
Title: Technology-driven innovation in Gulf Cooperation Council (GCC) countries : emerging research and opportunities / Anna Visvizi, Saad Bakry, Miltiadis D. Lytras, and Wadee Alhalabi, editors.
Description: Hershey, PA : Business Science Reference, [2019] | Includes bibliographical references.
Identifiers: LCCN 2018058918| ISBN 9781522590125 (hardcover) | ISBN 9781522590132 (softcover) | ISBN 9781522590149 (ebook)
Subjects: LCSH: Technological innovations--Persian Gulf Region. | Research--Economic aspects--Persian Gulf Region. | Industries--Persian Gulf Region. | Gulf Cooperation Council.
Classification: LCC T27.3.P35 T43 2019 | DDC 338/.0640953--dc23 LC record available at https://lccn.loc.gov/2018058918

This book is published in the IGI Global book series Practice, Progress, and Proficiency in Sustainability (PPPS) (ISSN: 2330-3271; eISSN: 2330-328X)

British Cataloguing in Publication Data
A Cataloguing in Publication record for this book is available from the British Library.

Practice, Progress, and Proficiency in Sustainability (PPPS) Book Series

ISSN:2330-3271
EISSN:2330-328X

Editor-in-Chief: Ayman Batisha, International Sustainability Institute, Egypt

MISSION

In a world where traditional business practices are reconsidered and economic activity is performed in a global context, new areas of economic developments are recognized as the key enablers of wealth and income production. This knowledge of information technologies provides infrastructures, systems, and services towards sustainable development.

The **Practices, Progress, and Proficiency in Sustainability (PPPS) Book Series** focuses on the local and global challenges, business opportunities, and societal needs surrounding international collaboration and sustainable development of technology. This series brings together academics, researchers, entrepreneurs, policy makers and government officers aiming to contribute to the progress and proficiency in sustainability.

COVERAGE

- Eco-Innovation
- Green Technology
- Innovation Networks
- Socio-Economic
- Knowledge clusters
- E-Development
- Outsourcing
- Global Business
- Technological learning
- Global Content and Knowledge Repositories

IGI Global is currently accepting manuscripts for publication within this series. To submit a proposal for a volume in this series, please contact our Acquisition Editors at Acquisitions@igi-global.com or visit: http://www.igi-global.com/publish/.

The Practice, Progress, and Proficiency in Sustainability (PPPS) Book Series (ISSN 2330-3271) is published by IGI Global, 701 E. Chocolate Avenue, Hershey, PA 17033-1240, USA, www.igi-global.com. This series is composed of titles available for purchase individually; each title is edited to be contextually exclusive from any other title within the series. For pricing and ordering information please visit http://www.igi-global.com/book-series/practice-progress-proficiency-sustainability/73810. Postmaster: Send all address changes to above address.

Titles in this Series

For a list of additional titles in this series, please visit:
https://www.igi-global.com/book-series/practice-progress-proficiency-sustainability/73810

Bioeconomical Solutions and Investments in Sustainable City Devlopment
José G. Vargas-Hernández (University of Guadalajara, Mexico) and Justyna Anna Zdunek-Wielgołaska (Warsaw University of Technology, Poland)
Engineering Science Reference • ©2019 • 300pp • H/C (ISBN: 9781522579588) • US $195.00

Optimizing the Use of Farm Waste and Non-Farm Waste to Increase Productivity and Food...
Leighton Naraine (Clarence Fitzroy Bryant College, St. Kitts-Nevis)
Engineering Science Reference • ©2019 • 223pp • H/C (ISBN: 9781522579342) • US $155.00

Handbook of Research on International Collaboration, Economic Development, and ...
Vasilii Erokhin (Harbin Engineering University, China) Tianming Gao (Harbin Engineering University, China) and Xiuhua Zhang (Harbin Engineering University, China)
Business Science Reference • ©2019 • 703pp • H/C (ISBN: 9781522569541) • US $235.00

Handbook of Research on Rural Sociology and Community Mobilization for Sustainable Growth
Qaiser Rafique Yasser (Centre for Rural Economy, Pakistan)
Information Science Reference • ©2019 • 393pp • H/C (ISBN: 9781522571582) • US $245.00

Impacts of Violent Conflicts on Resource Control and Sustainability
Esther Akumbo Nyam (Plateau State University, Nigeria) and Asiru Hameed Tunde (Umaru Musa Yar'adua University, Nigeria)
Engineering Science Reference • ©2019 • 218pp • H/C (ISBN: 9781522559870) • US $185.00

Practice and Progress in Social Design and Sustainability
Kin Wai Michael Siu (The Hong Kong Polytechnic University, Hong Kong) and Yin Lin Wong (The Hong Kong Polytechnic University, Hong Kong)
Information Science Reference • ©2019 • 330pp • H/C (ISBN: 9781522541837) • US $195.00

Handbook of Research on Methods and Tools for Assessing Cultural Landscape Adaptation
Isabel de Sousa Rosa (University of Lisbon, Portugal) Joana Corte Lopes (University of Lisbon, Portugal) Ricardo Ribeiro (University of Lisbon, Portugal) and Ana Mendes (University of Lisbon, Portugal)
Engineering Science Reference • ©2018 • 524pp • H/C (ISBN: 9781522541868) • US $265.00

For an entire list of titles in this series, please visit:
https://www.igi-global.com/book-series/practice-progress-proficiency-sustainability/73810

Table of Contents

Preface

The objective of this edited volume is twofold: (1) to showcase the state-of-the-art approaches to and successful applications of technology-driven innovation (as it is enabled by streamline, converging and emerging technologies) in the GCC countries and (2) to outline how new, advanced, international business, business innovation models, innovation and adaptive strategies can expand the economic- and social-sustainability frontiers in the GCC region. Consequently, chapters are sought that address these issues and – through novel, conceptually sound and empirically-focused approaches – extend our knowledge and understanding of the functioning and evolution of the global economy, including its regional and national domains.

The ongoing economic development and differentiation of the economic and social sustainability models in the GCC countries require that sophisticated approaches to the development of business models, entrepreneurship capabilities and skills and competencies are needed. These need to be anchored in advanced information and communication technologies.

This edited volume promotes an integrative, interdisciplinary and multidisciplinary approach to the analysis of processes as complex as those shaping contemporary developments in the GCC region, many of which are founded on bold cross-sectoral innovation strategies. To this end, chapters in this volume bring together the following conceptual, empirical and policy-oriented considerations:

- Skills and competencies required for a technology driven innovation era in the GCC countries;
- Strategies advancing digital transformation anchored in competitive areas and acquisition of new markets;
- Advanced computing approaches and information systems and their impact on international business leadership and business model innovation;

- Innovation networks and their role in fostering sustainable socio-economic growth and development at local, national, regional and global levels.

Contributions exploring diverse issues delineated by the above-mentioned research-areas, will seek to present the outcomes of the scientific inquiry in a manner conducive to their use for active learning. In this manner, this edited volume serves as a very much sought-after reference edition for postgraduate studies and executive seminars.

The economic development and the differentiation of the Sustainability Economic Model in Gulf Cooperation Council Countries will require sophisticated approaches for the development of business models, entrepreneurship capabilities and skills and competencies building anchored to advanced information and communication technologies. for enhanced learning contexts with enriched digital media elements. New learning strategies

The evolution of streamline and emerging technologies poses critical challenges to management. In the context of high performing economies, it is thriving to invest and to capitalize on sophisticated planning and implementation of technology driven innovation. The subject area of this book meets the cross-section of three disciplines:

- Innovation Management
- Emerging Technologies
- International Business

The convergence of these areas directly leads to a new perception of Innovative International Business, where the main emphasis is paid on the development of strategic competitive advantage built on the top of local skills and competencies with international oriented market strategies. This is somehow a new definition for the digital transformation which is the top priorities of many economies in the Gulf Cooperation Council Countries.

The focus of this edition is on the discussion of a 3-tier approach to digital transformation which is organized around the following pillars:

- Enhancement and Strategic Development of Skills and Competencies for the required Digital Transformation
- Development of a Strategic Plan for the Digital Transformation anchored to competitive areas and new market expansion

- Advanced computing approaches and systems for international business leadership
- Building of International Innovation Networks

SUBJECT AREAS

Sustainable economic growth, social capital, international business, technology driven innovation, are the key concepts that define the topics discussed in this volume.

This edited volume explores a four-tier approach to digital transformation, including the following pillars:

- Enhancement and Strategic Development of Skills and Competencies for the required Digital Transformation
- Development of a Strategic Plan for the Digital Transformation anchored to competitive areas and new market expansion
- Advanced computing approaches and systems for international business leadership
- Advancement of innovation networks at local, national, regional and global levels;

TARGET AUDIENCE

The audience of this edited volume consists of students, educators, international business experts, innovation experts, computer engineers, scientists, and non-technical readers who are interested in innovation, entrepreneurship, and international business not only in the GCC region but also globally. It can be used as a reference editions for all those interested in Innovation, Emerging Technologies and International Business as well as for textbook for advanced and focused Innovation and International Business courses.

TOPICS

Topics of interest include, but are not limited to, the following scope:

Strategic Objective 1: New Perspectives on Innovation and International Business

- Sustainability Research
- Smart Cities Research
- High Tech Industries Innovation
- Business Models for e-innovation
- Electronic Marketplaces
- Models and theories of International Business
- Global Innovation Networks

Strategic Objective 2: Skills and Competencies Management for Technology-Driven Innovation

- Active Learning Strategies
- Competencies and Skills assessment
- Job Profiling
- Matching Skills algorithms and systems
- Co-Labs for international collaboration
- Distributed Competencies schemas
- Cloud platforms for hiring and job flexibility
- Computer Science and International Business Programs
- Teaching Innovation in Post Graduate programs
- Academia Industry Collaboration

Strategic Objective 3: Digital Transformation Through Innovation

- Sectoral Studies in Education, Industry, Bioinformatics, Tourism, Services Industry, Logistics, Commerce, Retailing, E-services and Applications, Mobile Systems
- Sectoral Studies in Healthcare and Biotechnology,
- Sectoral Studies in Robotics, Nanotechnology, Green Energy, Green Computing
- Sectoral Studies on Public Administration and E-Government

Strategic Objective 4: Emerging Technologies Focus

- The Internet of Everything (IoE) Campus Infrastructure through Analytics Deployment
- Architecting University, Library, and Research Digital Repositories
- e-research, Data Curation, Management of Scholarly Identity
- Innovations for Improving Student Retention through Predictive Modeling
- Innovative Solutions for Student-retention Models, KPIs, Data dashboards, and Mobile Alerts to Identify At-risk Students
- Case Studies of Ethical Use of Student Data for Learning Analytics
- Data-driven Learning and Assessment
- Data Science with the aim of Learner Profiling
- Novel approaches to Analyze Individual Student Interactions in Online Learning Activities
- Integrating Analytics with Online Texts, Courseware, and Learning Environments to Measure Student Progress and Interaction
- Data Analytics Platform for Detailed Reporting, Assessment, and Collaboration
- Analytical Models on Students Data on Library use, Attendance, and Grades
- Intelligent Student Progress Dashboard
- Cloud-based audio and video creation tools to capture important human gestures, including voice, eye contact, and body language, which all foster an unspoken connection with learners
- Predictive Analytics Reporting (PAR) Framework, Learning, and Transfer
- Analytics, Business Intelligence, and Data Management for Higher Education
- Cutting-edge Asynchronous and Synchronous Tools to Advance the State of Online Learning
- Data-driven Evidence of Effective Blended Learning
- Stimulated Social Activities and Critical Thinking within an Online Environment
- The Effect of Analytics on Learning Outcomes and Advanced Learning Tasks

- Case Studies of Personalized Learning Using Analytics: range of educational programs, learning experiences, instructional approaches, and academic-support strategies intended to address the specific learning needs, interests, aspirations, or cultural backgrounds of individual students.

Strategic Objective 5: Policy Making for Technology-Driven Innovation

- Strategic Studies
- Roadmaps
- Actions and Guidelines
- Best Practices and Lessons Learnt

ORGANIZATION OF THE BOOK

The book is the first in a series of three books planned in order to cover the five strategic objectives analyzed in the previous section. It is organized into eight chapters.

A brief overview of each of the chapters follows:

Section 1: Drivers of Innovation in the Gulf Cooperation Council Countries – The International Economic Outlook

Chapter 1 – Social Capital: A Moderating Variable to Innovation
Chapter 2 – Technology-Driven Innovative Municipalities: The Case of Dubai Municipality
Chapter 3 – A Bibliometric Perspective on Technology-Driven Innovation in the Gulf Cooperation Council (GCC) Countries in Relation to Its Transformative Impact on International Business
Chapter 4 – A Bibliometric Analysis of the Technological Innovation in China

Section 2: Emerging Technologies and the Prospect of International Collaboration

Chapter 5 – Blockchain Technologically Driven Innovation at the Gulf Cooperation Council Countries: An Overview

Chapter 6 – Technologically Driven Legal Framework of Blockchain and Cryptocurrencies
Chapter 7 – Profile Analysis of Research Output in Biotechnology in Turkey

Chapter 1 discusses social capital as a moderating variable to innovation. Social capital is the foundation of innovative capabilities. There's no denying the fact that innovation is nothing but a collaborative effort of different actors in partnerships formed between universities, industries, and governments, and social capital plays an integral role in all the stages of the innovation process, from developing creative ideas to commercialization of those ideas. It's due to social capital that communication, dispersal of information, and sharing knowledge is done effortlessly. Thus, investing in the development and management of social capital is the most crucial thing for developing a collection of innovative capabilities and gaining the potential to use those capabilities in order to achieve competitive advantage in the market.

Chapter 2 is devoted to the technology-driven innovative municipalities: the case of Dubai municipality. The chapter sheds light on the linkages between technology-driven innovation, innovative capabilities, activities, and their impact on innovation performance in Dubai Municipality. The methodology adopted is qualitative research using various sources of data, including reports, academic papers, and a survey of visitors and residents. The chapter aims to depict the relationship among a set of elements that foster innovation in the urban context, which is an integral part of an overarching strategy for mainstreaming public sector innovation. Leadership, culture and infrastructure are critical determinants for an innovative municipality. A set of policy recommendations concludes the paper to enhance innovation performance and sustainable innovation in cities by investing in digital transformation, smart infrastructure, and e-governance.

Chapter 3 delivers a study on a bibliometric perspective on technology-driven innovation in the Gulf Cooperation Council (GCC) countries in relation to its transformative impact. This chapter presents a novel scientific research landscape of the Gulf Cooperation Council (GCC) in order to access the research productivity, scholarly impact and international collaborations across all GCC countries over the time period of 2008–2018, using the Scopus database. While we observe a significant increase in investing the resources in GCC world in order to build research infrastructure to make shift from only oil producing countries to knowledge-based economies, not enough efforts have been done to measure the outcomes of these investments.

The study explores different thematic areas that have evolved over a decade related to TDI, top scientists, and top institutions in GCC. Further, the study investigates international collaboration networks with-in or outside GCC to better understand the venues of knowledge exchange with GCC countries. It is expected that the findings of this case study would provide an insight to the research landscape of the GCC and useful information to the scientific community as well as to the technology and innovation policy makers.

Chapter 4 summarizes authors' contribution related to a bibliometric analysis of the technological innovation in china. It is used in order to compare the focus of research between GCC and China regions. As we have explained we have delivered in parallel one more book on China's Foreign Business.

This study uses direct citation network analysis and bibliometric analysis approach to explore the overall development of China's technological innovation research field, which can be classified into 5 groups. Group 1's main research topic is the interaction between technology innovation agents including state, local university research institutions, local companies and foreign companies. Among them, the state as a policy maker and a resource distributor, plays a pivotal role in the development of China's technological innovation capability. Group 2 explores the relationship between relationship network and the ability to build technological innovation. Scholars explores from different relationship network, including personal business relationship and political relationship of the company's managers and the industry cluster effect, on how to establish a knowledge exchange cooperative relationship. Most of the papers in Group 3 discuss the technological innovation in China's photovoltaic and wind turbine industry, trying to explain why China's grown by leaps and bounds in these two industries and the innovation challenges encountered. The papers in Group 4 discuss China's low-carbon clean energy technology. Several papers explore the flow and dissemination of its innovation capabilities by looking at the application of relevant patents at national or regional level. As there's direct relationship between technology innovation and R&D, so the papers under Group 5 investigate issues around R&D, such as management, unit establishment and the impact of cultural differences etc. Besides, R&D capabilities, product competitiveness and intellectual property rights protection are also important areas being reviewed. This study uses direct citation network approach to determine major themes of the technological innovation in China. By using the number of internal citations among the papers, we identified core papers under each development group. Researchers can directly refer to the core papers list and read the relevant papers in order to gain insight into the development of specific group, and explore research topics that are of particular interest to them, thus saving them a lot of time to filter out useful information from irrelevant papers.

Chapter 5 tries to determine blockchain technologically driven innovation at the Gulf Cooperation Council Countries. This chapter delves into the innovation of blockchain technology, and its benefits for advancing the economic growth in the Gulf Cooperation Council (GCC) countries. Blockchain technology provides the security and reliability that is required in the GCC region for enhancing the technology acceptance level of its population. Apart from providing security, blockchain technology offers financial efficiency, cost reduction, and the elimination of operational risks. Blockchain technology is equally beneficial for nonfinancial industries, such as the healthcare and education industry of the GCC region. The digital revolution based on blockchain technology is beneficial for GCC countries to shift their economic dependence from oil to innovations driven by technology. This is simply will move the GCC to develop more smart cities as the GCC region has an ample amount of financial and human resources that are required for successful integration of blockchain technology in its numerous industries.

Chapter 6 addresses the requirements of technologically driven legal framework of blockchain and cryptocurrencies. Blockchain technology refers to a digital, immutable, distributed ledger that registers completed transactions in a well-ordered manner and near real time. Blockchain security creates a decentralized environment that bars any third-party organization from controlling the cryptographically validated transactions and data. Blockchain technology fosters business innovation by creating a peer-to-peer networking which prevents one central server from accessing as well as processing data belonging to all companies in the network. Cryptocurrency can be defined as a digital asset built to facilitate completed transactions using cryptography. It helps in providing protection to the completed transactions and controlling the creation of additional units of the currency. In the recent years, the application of blockchain technology has been associated with governance. Blockchain governance has been applied in different field, for example it can be used to create permanent laws that cannot be violated by any third party.

Chapter 7 provides a study about a profile analysis of research output in biotechnology in Turkey. The aim of this study is to reveal the research profile and intellectual accumulation of Turkey in the field of biotechnology by using bibliometric methods for the period 1973-2016. To this end, the period studied in this framework is divided into two sub-periods (1973-2003 and 2004-2016). Thus, it was aimed to analyze the effects of the Vision 2023 strategy document, which appeared in the 2000s and which considers biotechnology as one of the strategic technologies to be developed as a priority. Using the dataset created from the scanned records in the ISI Web of Science database, Turkey's publication trends, collaboration patterns, preferred journals and research subject areas were identified. Despite the improvement observed in publication performance and co-operation for the 2004-2006 sub-period, Turkey has lagged behind countries such as Brazil, India and China, especially in terms of publication performance.

Acknowledgment

We dedicate our work to all scholars, academics, researchers, policy makers and industry leaders in the Gulf Cooperation Council Countries that envision social inclusive sustainable economic growth for all citizens, and seek to build bridges of peace and collaboration.

Also we dedicate our work to our collaboration as an example of mutual respect, friendship and trust between European and Arab Scholars.

Section 1

Drivers of Innovation in the Gulf Cooperation Council Countries:

The International Economic Outlook

Chapter 1
Social Capital:
A Moderating Variable to Innovation

Haytham Abduljawad
Knowledge to Power Consulting, USA

ABSTRACT

Social capital is the foundation of innovative capabilities. There's no denying the fact that innovation is nothing but a collaborative effort of different actors in partnerships formed between universities, industries, and governments, and social capital plays an integral role in all the stages of the innovation process, from developing creative ideas to commercialization of those ideas. It is due to social capital that communication, dispersal of information, and sharing knowledge is done effortlessly. Thus, investing in the development and management of social capital is the most crucial thing for developing a collection of innovative capabilities and gaining the potential to use those capabilities in order to achieve competitive advantage in the market.

INTRODUCTION

The century we're living in is full of examples pertaining to driving innovation. A number of new technologies, products, and services have surfaced due to successful innovation. In fact, whole new industries have emerged because of it. But still the need for innovation in businesses is quite intense, especially in developing countries, and more so in the Gulf Corporation Council (GCC) region.

DOI: 10.4018/978-1-5225-9012-5.ch001

The need exists because developing countries are still seeking economic growth and diversification. In the GCC countries, diversification is rather more needed than growth. Statistics show that GCC countries experience impressive growth. However, they lack diversification to sustain their growth and quality of life, especially when their current growth depends largely on global and regional geopolitical environment.

Innovation paves the way for economic growth and diversification. Innovation is the process that gathers information from the maturing industries and transforms them into new processes, products, and services to meet the ever-evolving market needs (Lalkaka, 2002). Sure, innovation gives an organization a competitive advantage, which is indeed conducive to business growth, but there's another reason why developing countries crave innovation. Developing countries need to innovate because innovation leads to the development of human capital. That is perhaps the least cited reason why developing countries must innovate. However, there's an obstacle that prevents developing countries from driving innovation.

The knowledge composed by a diverse workforce employed in an organization tends to accelerate the innovation process (Omerzel and Rune, 2011). The mere communication carried out between the individuals creates knowledge that is vital for the development of quantifiable innovations. This seemingly easy and natural process poses a challenge of enormous proportions for the nations with developing economies because developing nations, such as those in the GCC region, rely heavily on foreign workforces. Thus, all the strategies pertaining to knowledge management fail to anticipate the knowledge walkout syndrome – a pattern that is commonly observed in the GCC region and is the basic reason why developing countries to this date remain incapable of creating tacit knowledge, which cannot be quantified easily, but it's the most important prerequisite for innovation.

In the GCC region, a major chunk of tacit knowledge exists in the minds of expatriates who are working for various Multinational Corporations (MNCs) instead of within the national workforce of the region that is hardly ever going to transfer its business operations overseas. The problem arises when the semi-permanent or short-term employment contracts of these expatriates expire and they take the tacit knowledge capable of driving innovation with them when they leave the region. This phenomenon is known as knowledge walkout, and it's a huge challenge that comes in the way of developing nations in the GCC region trying to cultivate innovation. However, I argue that the real problem lies within the corporate culture that does not incentivize the sharing and conversion of tacit knowledge amongst their employees. That problem is compounded by the fact that most GCC corporations are a mix of ethnic cultures that have different interpretation of knowledge sharing. When cultural and ethnic norms overpower an organization's culture and norms, both tacit knowledge and the innovation process are negatively impacted.

Tacit knowledge is not the only prerequisite required for driving innovation successfully. Other than tacit knowledge, there are two essential prerequisites for innovation – the first is absorptive capacity, and the second is innovative capacity. Absorptive capacity is best described as an organization's potential to learn, understand, and use external knowledge (Cohen and Levinthal, 1989). Innovative capacity is the ability to innovate (Saviotti, 1998). It is the sum of all individual and organizational capabilities, policies, and infrastructure that support the innovation process by facilitating knowledge production.

In order to initiate the innovation process, it's essential for the governments of developing countries in the GCC region to support the concerned institutions and collaborate with them to formulate strategies that help build knowledge stocks continuously by capturing, transferring, and disseminating knowledge for national growth and development. These knowledge stocks are intellectual capital that is the most important asset of an organization and the key driver of innovation.

The education system and universities have a significant role to play in knowledge creation when it comes to driving innovation in a nation. The more scientists, academics, and engineers there are in a nation, the easier it will be to create knowledge. Thus, governments of developing countries must make policies geared towards improving the national education system. Moreover, significant support for universities and research work from the government is required to fasten the pace of the innovation process. That's where University Industry Government (UIG) partnerships come into play. These strategic partnerships between universities, industries, and governments facilitate the development of advanced technology and promote the creation and sharing of knowledge.

Given that sharing knowledge is important in order to enable the innovation process to progress, making new connections on both individual and organization level becomes all the more important. And making new formal and informal connections is not possible without social and professional interaction. This brings us to an essential moderating component of innovation, i.e. social capital. Bourdieu (1986, p248) defines Social Capital as "the aggregate of actual or potential resources which are linked to the possession of a durable network of institutionalized relationships of mutual acquaintance and recognition."

Social interaction is considered an integral component of the innovation process, especially in the knowledge-based economies (Landry, Amara and Lamari, 2000). Innovation is a process that introduces a new product or service but not without involving both formal and informal interactions and relationships within and between organizations. This network of communication and relationships in the innovation process promotes another type of capital that stimulates and sustains knowledge management. This capital is called social capital. A high level of social capital contributes to innovation by reducing transaction costs, search and information costs, decision costs, bargaining costs between the different actors involved in the innovation process.

There are three core elements that define social capital – trust, civic norms, and reciprocity (Putnam, 1995). Relationships built on these three core elements build the much-needed knowledge to drive innovation. That means innovation does not only depend on human capital and the accumulation of research and development (which are undoubtedly the pivotal drivers of innovation), but is also powered by the accumulation of social capital.

This chapter is geared towards binding together the two concepts, innovation and social capital, and highlighting the importance of social capital in the process of driving innovation, especially in the developing countries of the GCC region.

THE INNOVATION PROCESS

The innovation process comprises of four fundamental steps, with creativity being the heart of the whole process. The first step towards innovation is idea generation and it requires individual and organizational creativity. The collected ideas are then assessed and filtered formally. In the second stage of the innovation process, the selected ideas are critiqued in a research study which helps to determine the feasibility and commercial viability of the idea, taking into consideration both the quantitative and qualitative validation. The basic aim of this feasibility assessment is to evaluate whether or not the idea is legally, technically, and environmentally possible to develop further. If the outcome of the research qualifies the idea, work towards further development of the idea will begin by building concept and optimizing it. The last step of the process involves implementing the idea, making sure it's commercial enough to reap profits.

The entire process of innovation is about turning creative ideas into intellectual capital and investing that capital in commercial, philanthropic, or educational products and services that help make a nation productive and sustainable. Each step of the innovation process involves relevant knowledge. However, there's something more important involved in the process – determined strategic management that promotes a culture of creativity, manages a portfolio of diverse ideas, gathers organizational resources, and allocates financial resources.

The secret to the success of the innovation process is the contribution of a nation's scientists, academics, and engineers. These are the individuals who power and support the research work involved in the innovation process. It's safe to say that universities play a pivotal role in producing the essential human capital that drives innovation.

Knowledge as a Strategic Resource for Innovation

Knowledge is an essential intangible asset needed for driving innovation in a nation. It's a crucial and strategic resource that paves the way for differentiation and competitiveness. The way knowledge and information are created, shared, and used in an organization is a whole process known as knowledge management. And this process plays an integral role in helping an organization achieve its objectives. In short, knowledge management is all about making the best use of the created and shared knowledge in order to gain competitive advantage, improved performance, and innovation.

The Explicit and Tacit Knowledge

Knowledge can be bifurcated into two types – explicit and tacit knowledge. You can easily describe or communicate explicit knowledge, code it, and access it, while tacit knowledge, owing to its implicit nature, is more difficult to quantify and teach. It is internalized knowledge that an individual possesses but is not consciously aware of such knowledge. For instance, there is no prescribed way to teach complex social skills, like how to become a good leader or how to instill good leadership qualities in someone. You can only become a good leader with time and sufficient experience.

The innovation process develops fresh tacit knowledge, which is essential for gaining a competitive advantage, considering it's almost impossible to imitate it. In order to convert this new tacit knowledge into explicit knowledge, both formal and informal social interactions are important. This knowledge conversion process is known as “externalization”. It involves using a variety of collaboration systems, like online discussion database, blogs, forums, and wikis. It transforms the knowledge in people’s minds into documented structured forms.

Developing a Culture That Promotes Innovation

All the steps in the innovation process heavily rely on gathering relevant knowledge – from gathering creative ideas and identifying opportunities to the production of profitable products and services. Organizations that collaborate with one another develop innovative capacity by facilitating the flow of knowledge between the two organizations. Absorptive capacity is the prerequisite for innovative capacity. This means organizations must have the potential to learn, understand, and apply the knowledge. Every individual has a role to play. From students and researchers to experienced industry professionals, everyone has a key role to play in an attempt to develop an organization’s innovative capacities to produce unique, modern, and innovative products and services.

In order to enhance an organization’s innovative capacities, it’s important to develop a culture that encourages taking risk and learning from failures. In multicultural organizations in the GCC region, employees are risk averts because they want to avoid failure. The lack in trust for forgiveness in organizations, limit the social interactions for knowledge conversion to a minimum. This weakness needs to be addressed in order to develop a culture that promotes innovation.

Developing National Innovation System

The National Innovation System (NIS) stimulates the exchange of knowledge and fuels the innovation process among individuals, organizations, and institutions (Bontis, 1998). Developing National Innovation System is extremely important for initiating the innovation process on the national level.

The innovation system theory states that innovation and the development of technology is the result of complex relationships between the actors involved in the innovation process, such as universities, industries, and government research institutes (Lundvall, 1992). The higher the number of such relationships in a nation, the stronger the National Innovation System will be. Governments play a pivotal role in developing National Innovation System because they are in a better position to initiate funding interactions and to create partnerships with industries and universities. These UIG partnerships boost the environment that supports innovation.

Top-Down and Bottom-Up Approaches to Innovation

A number of industrial organizations and governments have employed a top-down approach to fuel innovation in the past. This traditional approach for stimulating innovation does have some merits:

- It makes sure that all the strategic objectives of an organization are aligned with one another.
- It defines and supports an innovation culture.

But the aforementioned merits do not mean that a top-down approach is flawless. This approach can be overly bureaucratic and might cause significant delays in achieving results. A top-down approach also fails to motivate the key stakeholders in the innovation process, like students, researchers, faculty, employees, and middle management. Given that human capital is a crucial asset required in the innovation process, the lack of importance given to human resource in a top-down approach is a considerable weakness in it. Adopting an approach that overlooks the significance of human capital would result in the ineffective participation of knowledge makers.

Alternatively, a bottom-up approach is a more democratic approach to innovation. This approach helps initiate innovation at the grassroots level and is managed with the help of each organizational level's participation. A bottom-up approach fosters creativity and promotes both individual and group thinking. It also supports open communication through and through. Open communication and a bottom-up knowledge management are not enough to drive innovation in an organization, or in a nation for that matter. Building an environment that is conducive to innovation requires both formal and informal inter-organizational partnerships, which are developed and fostered with the help of professional and social networking. A professional network is responsible for knowledge exchange that expands the absorptive capacity of an organization.

The Need for Innovation in Developing Nations

Developing nations need to learn valuable lessons from developed nations that can provide a range of examples of how an innovation management system can facilitate sustainable economies. Driving innovation in a developing country depends on international sources of knowledge. Mainly, these sources comprise multinational corporations (MNCs) that are operating in a host developing country.

The factors that power the national innovation performance in a developing country include partnerships, collaborations, and joint ventures with MNCs. It won't be wrong to state that the absorptive capacity of a developing nation usually relies upon its ability to take advantage of the vast knowledge of MNCs working in the country. Absorbing external knowledge effectively compensates for the lack of human and financial resources in a developing country. But, on the other hand, it also challenges the growth and development of national human capital and knowledge exchange that is necessary for the transformation of a knowledge-based economy.

The only way to overcome these challenges is to foster collaborative partnerships between governments and educational institutions of developing nations to build a robust national innovation system. Moreover, governments of developing countries must offer support and encourage education and extensive research to propel the national innovation process. With the help of knowledge created through university – industry – government (UIG) partnerships, a developing country can look forward to advancing its skills, technologies, products, and services.

Key Factors That Impact Innovation

Innovation is generally measured by various inputs, like the intensity and extent of research and development, and outputs, including new intellectual properties and patents. When we talk about national innovation measurements in the developing countries of the GCC region, we notice that innovation is measured through the expenditures of research and development and intermediate outputs, including new publications and patents. This is somewhat a product-oriented approach that ignores services as a major output of the innovation process. Fortunately, the latest models devised for the measurement of innovation have taken services into consideration as well. The modern models have not only included services but also other positive outcomes, like institutional linkages, socioeconomic development, and the improved positioning of countries within the global competitive market.

Today, innovation is impacted by the following factors:

- **Strategic Planning and Management:** One major output of an organization's strategic planning is an innovation strategy. An innovation strategy is all about identifying, selecting, and implementing an innovative idea. The strategic planning of an organization includes the vision and objectives set for driving innovation. It's the detailed road map that talks about what needs to be innovated, why it needs to be innovated, in what part of an organization's value chain must innovation take place, and the expected duration of time required for driving innovation.
- **Knowledge Management:** Innovation is all about creating and applying knowledge, especially in the initial stage of the innovation process where ideas are being generated to solve a problem. In this phase of the process, the knowledge that is created needs to be articulated. This is where knowledge management plays a vital role in identifying and disseminating viable ideas. This stage requires a high-level collaboration at both internal and external levels in order to create knowledge and make it flow between various stakeholders. If truth be told, knowledge is created at every stage of the innovation process, which makes knowledge management all the more important for driving innovation.
- **The Culture and Structure of an Organization:** It's possible that the current or existing culture and structure of an organization are not fit for driving innovation within it. In that case, managers need to modify the organization culture and structure to make it compatible with the innovation process. Adapting the organization culture and structure is what guides stakeholders on how to innovate and how to measure innovation performance. Most of the times, it's a flatter or more democratic organizational structure that is compatible with the innovation process because such business structures are not burdened by bureaucratic pressures. Organizations with deeper and broader structures can also drive innovation effectively and efficiently. But it's not just the type of organizational structure that matters. The success of the innovation process heavily depends on organizational culture also, which needs to be flexible and responsive to allow a change.

Adapting the organization structure and culture happens to be another challenge for the GCC countries because most organizations there that are involved in the innovation process are governmental or semi-governmental organizations that are widely known for their hierarchical organizational structures and bureaucratic cultures. What such organizations in the GCC region need to do is develop dedicated business units that are geared towards managing and facilitating the innovation process.

The Challenges for Developing Economies in the Innovation Process

Partnerships between universities and industries are crucial for developing human capital by means of education and training. Such collaborations are responsible for generating or acquiring knowledge from professional services, stimulating and focusing on research and development activities, and facilitating the applications of research and development outcomes. The challenge faced by developing economies, especially the ones that rely on limited natural resources, is that these economies fail to leverage the benefits associated with UIG partnerships. This is because the idea of fostering UIG partnerships is relatively new in developing countries, and because of this, a huge communication gap between industries and academia seems to exist. Both are unable to communicate their needs and challenges openly with one another. Moreover, the industries and academia fail to involve each other in their strategic planning initiatives. The lack of communication between the academia and industries of developing countries roots from the lack of confidence in each other's potential to address existing and future challenges pertaining to driving innovation. A unified vision of both the parties is vital for accruing benefits from UIG partnerships, which result in national economic growth.

Developed nations have demonstrated that UIG partnerships can prove to be mutually beneficial. But for UIG partnerships to be lucrative for all involved parties, it's extremely important that governments support these partnerships with their policies that are aimed towards facilitating long-term collaborations between universities and industries.

The Role for Universities to Play in the Innovation Process

Today, national innovation systems have become integrated with global innovation networks in various nations. That makes national innovation systems more dependent on foreign sources of knowledge. This trend is common in developing countries where local universities have joined hands with foreign universities of developed nations. This relationship promotes the absorption and adaptation of advanced knowledge and leads to the creation of novel ideas.

There is a significant difference between the roles played by universities in developing countries and in the developed countries, especially during the development and commercialization stages of the innovation process. For instance, universities in developed economies collaborate with industries in order to develop inventions, commercialize those inventions, and produce spinoffs subsequently. This is not the way these stages of the innovation process are carried out in the developing countries. In the GCC region, the countries do not have enough resources and infrastructure to carry out activities necessary for driving innovation, like developing or creating prototypes. Moreover, universities are not even supported by industries to market their inventions and earn a profit. Thus, the only thing developing countries have on which the effectiveness of their innovation activities depends is their connections and links with a variety of sources of information, technologies, knowledge, experience, human and financial resources.

The Role for Industries to Play in the Process of Innovation

Industries have a vital role to play in the innovation process that supports a knowledge-based economy. Business challenges are a perfect source for creating excellent and innovative ideas. These challenges are what motivate organizations to seek assistance from universities and form partnerships with the academia. Industries collaborate with universities in three ways – for research and development, education and training, and professional services. Education and training are two things that lead to the generation of ideas, which are eventually assessed through the means of extensive research. As far as professional services are concerned, these services also result in the creation of ideas, but these ideas are related to commercializing innovative products and services.

The Role for Governments to Play in the Process of Innovation

Whether the innovation process is adopted by organizations or by developing nations, networking, learning, and creating and sharing knowledge are the most important factors involved in it. There are a number of questions that must be raised before directing national efforts towards the development of frameworks necessary for learning, networking, and knowledge creation, and for exchanging that knowledge. Government policies provide immense support to the innovation process by formulating regulations pertaining to international trade, foreign direct investment, financial markets, labor markets, education, and general business environment with a special focus on SMEs.

When we talk about the government policies in the GCC region, we notice that government policies allow market monopoly, which reduces the competition within the market. This status quo must be challenged, and in order to do so, the governmental policies in the GCC region must aim to reduce anti-competitive markets. Such policies will effectively stimulate the innovation practices.

The Framework for Driving Innovation in the GCC Region

In order to drive innovation in the GCC region, the GCC countries should adhere to the following steps:

1. Assess the strengths, weaknesses, opportunities and threats (SWOT) of the national innovation systems in GCC countries and formulate innovation policies based on the SWOT analysis. Each GCC country has extensive experience in various industries. Owing to that, the focus of the policies should be on the industry in which each country excels. Moreover, the aim of the policies must be to enhance productivity in the existing value chain.
2. Create a regional human capital repository, to identify, assess and categorize competencies and develop a regional skilled workforce. The workforce will comprise all the key actors of the innovation process, including academicians and industry experts. Also, different industry sectors should be charged with the responsibility of updating the repository based on the current and future needs related to competency and skills.

3. Relax labor policies all over the region to make knowledge workers in the region more mobile. Doing so would increase the number of pivotal experts required in the innovation process and would promote knowledge diffusion.
4. Give meaningful incentives to the regional academia and make inter-industry collaborations to reduce redundancies, especially in the research and development initiatives. Also, aim for building innovation clusters all over the region.
5. Streamline a regional Intellectual Property Rights (IPR) policy to manage the high direct costs of copyrights and patents effectively and efficiently. Also, aim for encouraging foreign direct investments in the innovation process. A streamlined IPR policy must be geared towards minimizing discouraging factors that impact innovators, and copyrights and patent owners. Moreover, an IPR policy must not be applied uniformly across all industry sectors. It must be more responsive and should take into consideration the innovation lifecycle. Implementation of an effective regional IPR policy would lead to better allocation of resources, a reduction in the risk of duplication of work, and above all, a regional exchange of knowledge.
6. Increase competitiveness within the market in order to generate more innovative ideas for commercialization of products and services. Intense competition would lead to knowledge creation and performance improvement.

A UIG partnership can be a complex relationship between different institutions and organizations. There are numerous challenges involved in UIG partnerships that lead to conflicts. Challenges like considerable differences in opinions, point-of-views, interests, constraints, objectives, and incentives often lead to bigger and more serious conflicts, misunderstandings, and distrust. These conflicts are not only between the partners but also within them. As a result of these unsettling disputes and conflicts, the effectiveness of a UIG partnership ends up being compromised. Therefore, a UIG partnership should be managed by bridging the gap between different cultures. In fact, that must be the key consideration in every UIG partnership. The foundation of every UIG partnership is mutual trust (Barney and Hansen, 1994), and without it, it will fail.

Trust is a huge factor that affects innovation, and it is fundamental to the creation of social capital relations, which influence the effectiveness and efficiency of UIG partnerships and mitigates the challenges between universities and industry institutions (Tsai and Ghoshal, 1998). Besides trust, there are two other core elements of social capital – reciprocity and norms. A combination of these three elements can moderate a UIG partnership and make it successful by facilitating the transfer of knowledge.

EXPLORING AND UNDERSTANDING THE NUANCES OF SOCIAL CAPITAL

Social capital is not a new concept used for explaining the innovative process. The theorists of social capital have claimed that innovation can be propelled by social capital and that the high levels of social capital are extremely important for the societies to function effectively in a knowledge-based economy (Bao, 2008). In developing countries, social capital is mainly developed by multinational corporations because it's valuable for their success and for the development of national innovation process. But these organizations face a number of challenges while they create and diffuse knowledge within and between organizations. The common challenges faced by the MNCs in the developing countries that can be mitigated by social capital include:

- Employees being completely unaware about who has the knowledge that they seek.
- Employees not confident about the knowledge they possess
- Improperly aligned norms and lack of trust that affects knowledge sharing. This usually results from the kind of perception employees have about job security and their concerns pertaining to the authenticity and relevance of their knowledge.

The Dimensions of Social Capital

UIG partnerships are facilitated by social capital relations. These partnerships are based on social interactions between the different actors involved in them. There are three dimensions of social capital – cognitive, structural, and relational (Coleman, 1988).

Cognitive Dimension

The cognitive social capital relies on the shared values and culture that facilitate the understanding of appropriate ways to act in the social system. Also, many researchers have proved a positive correlation between organization performance and cognitive social capital with different cultures and norms shared by the members of a social network facilitating the exchange of knowledge (Lane, Salk and Lyles, 2001).

Cognitive social capital also has a profound effect on UIG partnerships in the GCC region. The academic institutions and industry organizations in the GCC region tend to differ in terms of their priorities, business models, cultural orientation, and philosophies. This is the reason why social and professional interactions are not very effective or beneficial to each partner if the partnership is established just for gaining exposure in the local and regional media. Such differences lead to lack of trust between the partners, making them lose faith over each other's capabilities.

In order to overcome this challenge, the first thing that organizations need to do is clearly state their values and norms to spare one another from any confusions and ambiguities in the future as the partnership progresses. Moreover, organizations must integrate their norms and values with their strategic management processes. As a result, partners will have to make efforts to align their values, missions, visions, and strategic objectives at the onset of the partnership.

Diversity

It's not just culture, values, and norms that affect cognitive social capital. Diversity plays an equally important role as well. The broad definitions of diversity encompass various differences among people, including their age, gender, ethnicity, and race (Esty, Griffin and Schorr-Hirsh, 1995). Organizations need to realize that there are various differences between people and groups and must implement a broader approach to cater to those differences and become inclusive organizations. There's no denying the fact that diversity is responsible for stirring great changes within an organization. It has the power to yield superior work productivity and competitive advantages.

Diversity in the GCC region is very high. Owing to the fact that the number of endogenous workforces in the GCC region is low, the said region is more diverse than any developed nation. Even though the impact of such high diversity on the development of social capital, especially trust, is not extensively studied in the region, the few studies suggest that diversity can affect social capital adversely if it is not managed properly. This turns out to be a major challenge for organizations in the GCC region. In order to overcome that challenge, organizations need to seek ways to manage diversity for the good. Policies must be formulated and leadership and managerial skills should be adapted to accommodate workplace diversity and provide a more multicultural work environment.

Norms

Norms are not to be confused with values. Organizational values are drafted so that employees can distinguish between what's desirable and proper and what's not. These values root from the culture of an organization. On the other hand, norms are like a social contract which consists of unwritten rules that are supposed to guide human behavior (Dakhli and de Clerq, 2004). It won't be wrong to state that norms are actually true reflections of organizational values.

In multicultural organizations, serious issues can arise if all the varying behaviors are not incorporated in organizational values. For maintaining the level of trustworthiness and the existence of effective norms in such organizations, it is highly important to develop a closed social network, which is designed to strongly interconnect and reinforce the relationships between different institutions and actors. The development of such a social network facilitates communication, strengthens trust, and results in mutual benefit. This is how an organization's social capital is strengthened.

Reciprocity

Repaying people for their kindness and affection that they have shown towards you is called reciprocity – the absence of which from social interactions leads to reduced cooperation, disappointment, and negative feelings. Reciprocity is like an informal framework for building human relations. Reciprocity plays an important role in UIG partnerships because the intention of individuals to exchange knowledge depends on the quality and quantity of exchange made with their counterparts (Molm, 2003). This means that individuals will feel reluctant to exchange knowledge if the exchange is not favorable for them or if they're not getting knowledge in return that benefits them.

Structural Dimension

This dimension of social capital is all about how relationships are created and configured (Hitt, Lee and Yucel, 2002). These relationships can be formal or informal, among organizations or individuals, etc. In short, the structural dimension is about who you reach and how you reach them. Obviously, social networks are the most common platforms used to reach people. These social networks can be either open or closed. Open social networks allow knowledge exchange to play a dominant role in UIG partnerships. The actors in UIG partnerships can explore knowledge resources in other actors outside an organization. On the other hand, closed social networks only involve actors within an organization.

It was mentioned earlier that how diverse workforces are in the GCC region. Organizations in that region have culturally and ethnically diverse workforces. This fact makes social networks crucial for their sustainable advantages. Moreover, organizations need to nourish and support the development of social networks if they want to create effective partnerships with others. In UIG partnerships, social networks are used for both developing and managing relationships for the exchange of knowledge and for converting knowledge into innovation. Thus, open social networks are required for creating knowledge and closed ones are required to assimilate it. Both types of social capital network structures are essential in UIG partnerships.

Relational Dimension

The relational dimension of social capital alleviates the noise that impacts the social interaction between the partners by building trust and making communication between partners easier (Fukuyama, 1995). This is the dimension of social capital where trust plays an integral role. According to this dimension, social capital is developed and fostered when individuals trust other partners to reciprocate appropriately and meet the expected obligations responsibly.

Trust

Trust is basically the belief that the results of an individual's intended actions will be appropriate from the point of view of another individual (Misztal, 1996). There are two types of trust – generalized and institutional trust. Simply put, generalized trust facilitates interaction and communication within a social context, and institutional trust facilitates the same in an organization. Organizations must institutionalize practices that build trust among individuals and teams both within and across their boundaries in order to capture, create, and exchange knowledge. When trust is developed, a lot of formal and bureaucratic processes are minimized, which is very important in the GCC region. The credibility of partners may not be untrustworthy in the GCC region, but the universities in the region lack trust in industry experience. This results in industry actors placing less confidence in the faculty members of universities and not considering them as reliable contributing partners.

Creating Social Capital with Multicultural Workforce is a challenge that is not limited to GCC region. Unplanned UIG partnerships only lead to further chaos and challenges. Such partnerships cause organizations to digress from their core objectives because multicultural UIG partners have varying and contradicting priorities, which impact the agenda of the partnership. In the GCC region, the diversity in working environments tends to affect social capital relations. The multicultural workforce creates several complexities in creating partnerships at many levels. Dealing with these complexities is the only way to allow social capital to create knowledge.

Firstly, multicultural individuals should be encouraged to share knowledge. Secondly, in a UIG partnership, organizations need to realize that the differences in goals and approaches when they aim to collaborate with one another. This would minimize the challenges in UIG partnerships and allow the relationships within UIG partnerships to create social capital that is beneficial for both the economy and society.

How Social Capital Contributes to Economic Growth

In a knowledge-based economy, social capital is an important prerequisite for driving innovation effectively and efficiently because it builds trust among social network members. There's no denying the fact that social capital has a positive effect on human capital and enhances communication and knowledge creation. This altogether impacts the economic growth of a nation.

The institutions involved in the innovation process expand their conventional roles and activities into those of other institutions. This is what makes them able to create and exchange knowledge. In order to carry out this expansion, the institutions have to develop social networks in environments and territories that are not even familiar to them and often conflict with their own norms and culture. When mutual trust, values, reciprocity, norms, and networks accelerate the development and exchange of knowledge, economic growth is inevitable.

CONCLUSION AND KEY TAKEAWAYS

This chapter is based on a conceptual framework designed as part of a case study research on University Industry Government (UIG) partnerships and their impact on national innovation systems (Abduljawad, 2013). The case study was conducted in 2013 in the state of Qatar, a member of the Gulf Cooperation Council Countries (GCC). The framework identified two factors that would moderate the stimulation of innovation in the region. The focus of this chapter is on UIG partnerships are new and involve complex processes in the GCC region. These partnerships have not been fully implemented, supported, or monitored. Thus, I selected a methodology relating to a qualitative, exploratory, case study to research a phenomenon that was new and unfamiliar (Creswell, 2007). The aim of the research was to explore UIG partnerships, determine how they contribute to the national innovation system, and whether it could be reasonably concluded or determined that these partnerships have the potential to stimulate innovation and accelerate the transformation of the state's national economy. The research was based on a methodology of naturalistic inquiry (Myers, 2008; Ponterotto, 2005), to explore and interpret the perspectives and experiences of research participants. The case study attempted to understand how does social capital affect UIG partnerships?

Social capital is the bedrock of innovative capabilities. It's true that innovation is primarily a collaborative effort and social capital plays the central role in developing creative ideas and managing the innovation process. It's because of social capital that communication, diffusion of information, and sharing and assimilating of knowledge is done smoothly and with fewer challenges. Thus, making investments in the development and management of social capital is therefore crucial for developing a wide range of innovative capabilities and gaining the flexibility to use these capabilities selectively in order to address constraints in the market and achieve competitive advantage.

However, the following impediments impact social capital and UIG partnerships in the GCC region:

- **Drifting Missions and Lack of Shared Visions:** This challenge is basically based on the lack of proactive leadership at the organizational level that can inspire and influence followers. The absence of quality leadership leads to great misunderstandings as a result of which organizations lose focus and direction. UIG partnerships require organizations to align their visions and have consensus among their employees on initiatives and activities that need to be taken.
- **Lack of Shared Values:** It was mentioned earlier that UIG partners are from different backgrounds. Owing to that fact, they have different values, behaviors, and norms as well. Thus, it is advised to arrange regular meetings, including kickoff meetings, so that partners from diverse backgrounds get a chance to have a dialogue about establishing and updating norms if there is a need to do so.

In order to make successful UIG partnerships and drive the innovation process effectively, the GCC governments need to remember the following:

- **UIG Partnerships are Built on Mutually Accepted Norms and Shared Visions:**The basic goal of UIG partnerships is to engage all the partners involved. This will lead to the generation of mutual trust. Thus, it's imperative for the organizations in the GCC region to develop leadership and management skills so that the entire workforce is united behind the visions and norms of these partnerships.
- **Derive Benefits from Diverse and Multicultural Workforces:** The diversity in the GCC region has a completely different meaning than in the diversity in the rest of the world. The diversity in other parts of the world might reflect on the changing world and dynamics of the marketplace, but in the GCC region, diversity is a matter of sustainability. In either case, diverse workforces are of great value to organizations. Rules must be bent if need be in order to incorporate more individuals from diverse backgrounds. This is what gives an organization a competitive advantage.
- **Recognize Individual Differences with the Help of Networking Skills:** Organizational leaders in the GCC region must accept and cater to the differences within their multicultural teams. Moreover, they should work towards leveraging those differences to derive the best outcomes from them.
- **Trust is the Part and Parcel of Social Capital:** UIG partnerships that are based on collaborations with people who have common goals,

visions, and norms tend to develop more fruitful collaborations out of their social network interactions. Trust among the partners facilitates knowledge sharing and stimulates the development and maintenance of both formal and informal social network interactions among the concerned stakeholders.

- **Norms Might be Unwritten but They Have an Undeniably Huge Impact on Corporate Culture:** It is indeed a difficult task to document all the norms that come with a multicultural workforce in the GCC region, but that doesn't mean norms lack in importance. It is believed that unwritten rules are worse than the ones that are poorly written. It is true only if they happen to contradict each other.

REFERENCES

Abduljawad, H. (2013). *University–industry–government partnership: State of Qatar transformation to a knowledge-based economy case study* (Doctoral dissertation). Bellevue University.

Barney, J. B., & Hansen, M. H. (1994). Trustworthiness as a source of competitive advantage. *Strategic Management Journal*, *15*(S1), 175–190. doi:10.1002mj.4250150912

Bontis, N. (1998). National intellectual capital index: The benchmarking of Arab countries. *Journal of Intellectual Capital*, *5*(1), 13–39. doi:10.1108/14691930410512905

Bourdieu, P. (1986). The Forms of Capital. In J. Richardson (Ed.), Handbook of Theory and Research for the Sociology of Education. New York: Greenwood Press.

Cohen, W., & Levinthal, D. (1989). Innovation and learning: The two faces of R&D. *Economic Journal (London)*, *99*(397), 569–596. doi:10.2307/2233763

Coleman, J. (1988). Social capital in the creation of human capital. *American Journal of Sociology*, *94*, S95–S120. doi:10.1086/228943

Creswell, J. (2007). *Qualitative inquiry & research design: Choosing among five approaches* (2nd ed.). Thousand Oaks, CA: Sage.

Dakhli, M., & de Clercq, D. (2004). Human capital, social capital, and innovation: A multi-country study. *Entrepreneurship and Regional Development*, *16*(2), 107–128. doi:10.1080/0898562041000 1677835

Esty, K., Griffin, R., & Schorr-Hirsh, M. (1995). *Workplace diversity. A manager's guide to solving problems and turning diversity into a competitive advantage*. Avon, MA: Adams Media Corporation.

Fukuyama, F. (1995). Social capital and the global economy. *Foreign Affairs*, *74*(5), 89–103. doi:10.2307/20047302

Hitt, M., Lee, H., & Yucel, E. (2002). The importance of social capital to the management of multinational enterprises: Relational networks among Asian and Western firms. *Asia Pacific Journal of Management*, *19*(2/3), 353–372. doi:10.1023/A:1016247920461

Lalkaka, R. (2002). Technology business incubators to help build an innovation-based economy. *Journal of Change Management*, *3*(2), 167–176. doi:10.1080/714042533

Landry, R., Amara, N., & Lamari, M. (2000). *Does Social Capital Determine Innovation? To What Extent?* Paper prepared for presentation at the 4th International Conference on Technology Policy and Innovation, Curitiba, Brazil. Retrieved from: http://www.inovacijos.lt/inopagalba/cms/114lt.pdf

Lane, P., Salk, J., & Lyles, M. A. (2001). Absorptive capacity, learning and performance in international joint ventures. *Strategic Management Journal*, *22*(12), 1139–1161. doi:10.1002mj.206

List, F. (1841). *The national system of political economy*. London, UK: Longman.

Lundvall, B.-Å. (Ed.). (1992). *National Innovation Systems: Towards a Theory of Innovation and Interactive Learning*. London: Pinter.

Misztal, B. (1996). *Trust in modern societies*. Cambridge, UK: Policy Press.

Molm, L. (2003). Theoretical Comparisons of Forms of Exchange. *Sociological Theory*, *21*, 1–17.

Mowery, D., Oxley, J., & Silverman, B. (1996). Strategic alliances and inter–firm knowledge transfer. *Strategic Management Journal*, *17*(S2), 77–92. doi:10.1002mj.4250171108

Myers, M. D. (2008). *Qualitative Research in Business & Management*. London: Sage.

Omerzel, D. G., & Rune, E. G. (2011). Knowledge Resources and Competitive Advantage. *Managing Global Transitions*, *9*(4), 335–354.

Ponterotto, J. (2005). Qualitative Research in Counseling Psychology: A Primer on Research Paradigms and Philosophy of Science. *Journal of Counseling Psychology*, *52*(2), 126–136. doi:10.1037/0022-0167.52.2.126

Putnam, R. (1995). Bowling Alone: America's Declining Social Capital. *Journal of Democracy*, *6*(1), 65–78. doi:10.1353/jod.1995.0002

Saviotti, P. (1998). On the dynamics of appropriability of tacit and of codified knowledge. *Research Policy*, *26*(7–8), 843–856. doi:10.1016/S0048-7333(97)00066-8

Tsai, W., & Ghoshal, S. (1998). Social Capital and Value Creation: The Role of Intrafirm Networks. *Academy of Management Journal*, *4*(41), 464–476.

Chapter 2
Technology-Driven Innovative Municipalities:
The Case of Dubai Municipality

Ahmed O. El-Kholei
https://orcid.org/0000-0001-7445-3641
Arabian Gulf University, Bahrain

Odeh R. Al Jayyousi
Arabian Gulf University, Bahrain

ABSTRACT

The chapter sheds light on the linkages between technology-driven innovation, innovative capabilities, activities, and their impact on innovation performance in Dubai Municipality. The methodology adopted is qualitative research using various sources of data, including reports, academic papers, and a survey of visitors and residents. The chapter aims to depict the relationship among a set of elements that foster innovation in the urban context, which is an integral part of an overarching strategy for mainstreaming public sector innovation. Leadership, culture, and infrastructure are critical determinants for an innovative municipality. A set of policy recommendations concludes the chapter to enhance innovation performance and sustainable innovation in cities by investing in digital transformation, smart infrastructure, and e-governance.

DOI: 10.4018/978-1-5225-9012-5.ch002

INTRODUCTION

An innovative city is a smart, digital city that is globally connected and whose urban development is knowledge-based. The key to an innovative city is an innovative municipality that fosters models of innovation in the public sector. Also, it embodies models of open and user innovation.

Dubai is a member of the United Arab Emirates (UAE). It is located in flat sandy land covering 1,287.5 square kilometers. Dubai enjoys moderate winter; however, summer is hot and humid. According to the Statistics Centre of Dubai, the population of the emirate reached 3,171,723 in December 2018 ("Population," 2018). Trade, transportation, tourism, aviation, real estate, and financial services are the major economic sectors. The city is among the most expensive cities in the Middle East (Swan, 2014). Dubai has attracted investors, entrepreneurs, and skilled labor by developing mega-engineering projects, such as Burj Khalifa. In 2013, UAE nationals represented about 15% of the population of the emirate; probably for this reason, the male-female ratio is skewed ("Dubai population," n.d.).

In 1954, the ruler of Dubai founded Dubai Municipality (DM) with a cadre of seven employees undertaking simple tasks in cleaning the city. He then appointed 23 municipal council members to be responsible for health and architectural matters of the city, as well as to organize the construction and beautification of the city and provide constructive suggestions to the government. DM went through successive phases of development and growth. Today, DM is the leading institution that drives the development of Dubai. DM is one of the largest governmental institutions. It employs 11 thousand employees working in 34 departments under the following six sectors:

1. International Affairs & Partnership Sector.
2. General Support Sector.
3. Environment, Health and Safety Control Sector.
4. Environmental and Public Health Services Sector.
5. Planning and Engineering Sector.
6. Corporate Support Sector.

According to several reports, such as Jones Lang LaSalle Incorporated and The Business of Cities report issued in 2018, and the Global and World Cities Research Network's results published in 2016, Dubai has good connections to the world economy, as it provides services to the international economy.

These reports assure that Dubai is both an emerging mid-sized new world city that is competitive in specific markets.

The innovative municipal administration is among the factors that contributed to the evolution of Dubai as a global, digital and smart city. In 2017, the DM won eight awards and six honorary certificates from international and Arab institutions devoted to ideas and innovation. The initiatives that received awards and certificates include the Integrated System for Medical Waste Treatment in Dubai; Currency Charity Bank; Irrigation System with Solar Power on Roundabouts (Admin, 2018). These initiatives contribute to the city's sustainability by minimizing waste, promoting social justice, and conserving resources.

Visvizi, Lytras, Damiani, & Mathkour (2018) attempted to link technology-driven innovation on the one hand, and both social and economic sustainability. They argued that technology enablers lead to a smart city strategy, thus raising social wellbeing. Technology-innovation interface is clear in private sector companies. It is not the case in public sector institutions. The debate to which this chapter contribute is to highlight linkages and relationship between technology and innovation in the public sector.

The chapter reviews the concepts and models of innovation in the public sector and examines their relevance to Dubai's success story. Specifically, the chapter investigates the role of the three pillars for public sector innovation. First, the chapter examines innovation activities within DM, such as ideas that clients and staff members bring to the attention of responsible parties within the organization. Second, the chapter assesses the innovation capability that infrastructures and funding models manifest that can sustain the innovation activity of DM. Next, the chapter attempts to measure the impact of innovation on the performance of DM including outcome, efficiency, and effectiveness. The researchers are after concepts, including management and leadership. The chapter attempts to shed light on the role and impact of the external environment of DM, such as opportunities and challenges conducive for innovation. These include culture, incentives, as well as external networks and partnerships.

LITERATURE REVIEW

Urban development and innovation interconnect with national innovation systems (Godin, 2007). Knowledge hubs and education infrastructure including innovation labs, incubators, and accelerators help in bridging the innovation system and urban innovation that is central to forming a regional innovation system. However, socio-technical landscape, regime, and niche innovations underpin innovative cities (Geels & Schot, 2007). The interactions among strategy, institutions, and culture contribute to defining three types of knowledge, namely, the artistic, creative, and imaginative that in turn shapes the innovative city in space and time (Asheim, Isaksen, Moodysson, & Sotarauta, 2011). Urban innovation can be cultural, such as Athens, technological, such as Manchester, or organizational, such as London (Hall, 2001). They are facets of the transformation that the city experienced as a result of the transformations of the capitalist system (El-Kholei, 2018).

Besides, this notion is in line with the idea of path dependency that argues that each innovative city has a historical trajectory and cultural context (Simmie, 2005). In the new models of creative cities, there is a delegation of responsibility from nations to cities to ensure competitiveness. Hence, the metrics for measuring innovation capability is between cities, not nations.

Cities are regional nodes for innovations that contribute to economic agglomeration (Inkinen, 2015). A framework for spatial innovation included concepts, such as industrial districts and clusters, that operate in science parks, living labs, and technology parks (Anttiroiko, 2016; Kleiman & Hillard, 2016). Enablers for urban innovations include openness, infrastructure, and leadership. There is evidence for the linkages between urban form, connectivity, and knowledge creation to foster urban innovation (Simmie, 2005). Digital startups and small and medium enterprises in cities are primary drivers for economic growth. Also, urban living labs, crowdsourcing for generating new ideas, and collaborative and participatory innovation platforms constitute vital enablers for urban innovation. Creative industries are crucial for shaping a sustainable economic base for cities (Al-Jayyousi, 2004).

The emergence of smart and innovative cities is a manifestation of neoliberal urbanization, where the state plays the role of an entrepreneur and impacts the way urban space is imagined and produced. These new examples of cities are new informed governance models, as public-private partnerships. The innovative city model attempts to produce conditions for a spatially-enabled society through information and communication technology (ICT) platforms that are likely to induce radical and disruptive innovation. Indicators for a smart and innovative city include smart economy, mobility, environment, living, and governance (Al-Jayyousi, 2017).

Innovation is the process that entails a non-linear, dynamic, and multi-actors-sectoral interactions, which include developing and implementing new processes, products, and services to enhance efficiency, effectiveness or quality (Bankins, Denness, Kriz, & Molloy, 2017). Today, manufacturing, oil and gas production, and agriculture are not the only economic sectors that can spearhead economic growth. Other sectors, such as services (i.e., education, tourism, creative industries, logistics, and transportation), contribute to economic growth. Traditionally, research in the field of innovation focused on technological advances and developing new products. Recent research in the field of innovation has started to address public sector innovation and services innovation that is expanding into a diverse and cross-disciplinary body of knowledge, including economics, management, administration, and marketing (Randhawa & Scerri, 2015; Sanders Jones & Linderman, 2014).

Institutional inertia is among the barriers to innovation in public sector organizations. Research shows that, in many instances, new ideas from the people dealing with the organization or innovations from middle managers and frontline staff do not get through the bureaucracy (Hughes, Moore, & Kataria, 2011). Determinants and enablers for innovation include incentives and organizational culture (Moussa, McMurray, & Muenjohn, 2018). However, organizations are likely to face asymmetry of incentives where there is a tendency to punish unsuccessful innovations much more severely than reward successful ones (Teofilovic, 2002).

The models and frameworks that address public sector innovation examine relationships between innovation activities, innovation capacity, and conditions to measure the performance of the public sectors (Borins, 2016). Figure 1 illustrates the interactions between innovation activities, which include accessing, selecting, implementing, and diffusing innovative ideas and services. Besides, the innovation capability (which is part of the conditions of innovation) entails the management of innovation, leadership, incentives, and culture. The impact on performance includes enhancement in service and efficiency.

Individual innovation requires leadership to manage knowledge and cope with change and discontinuity (de Vries, Bekkers, & Tummers, 2016). Lynn (2013) argued that innovation might entail a radical and structural change in the business model, flow of knowledge, and service delivery. Besides, innovation entails changes in operations (Borins, 2001; Hughes et al., 2011),

Figure 1. Framework for innovation in public sector organizations
Source: adapted from Hughes et al., 2011

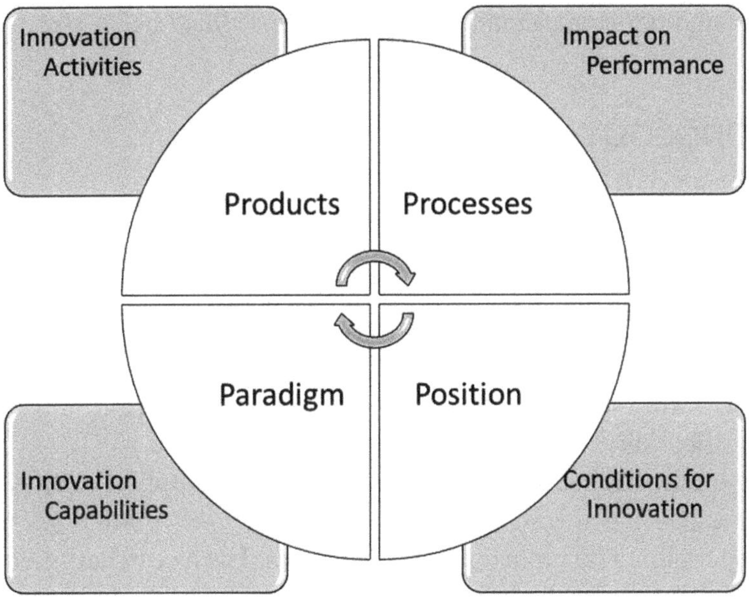

promoting public value (Asheim et al., 2011; Caulier-Grice & Mulgan, 2010), inducing discontinuous change (Arundel & Hollanders, 2011) and focus on implementation and diffusion of new processes (Gow, 2014), and formation of new relationships with users and service providers (Bankins et al., 2017). Moreover, innovation is linked to the abilities of individuals to generate new ideas and put them into practice (Hughes et al., 2011; Walker et al., 2008).

Also, the culture of innovation in the public sector can be enhanced by developing the core competencies, avoiding barriers to innovation, and developing conditions to foster innovation and performance (Kuhlmann & Rip, 2014).

There is evidence that innovation is among the drivers for modernizing public administration (Cucciniello & Nasi, 2014a, 2014b). Research has shown that innovation in the public sector may play an important role to enhance the performance of the public sectors (Kuhlmann, S., & Rip, 2014b; Turnheim & Geels, 2013).

Viewing innovation within space and time was manifested in many domains, such as organizational learning, innovative cities and municipalities, innovative regions, learning communities, and knowledge hubs (Al-Jayyousi,

2004; Yun, Jeong, & Yang, 2015). Dubai City represents a new business model for a creative city that wider-sector conditions, innovation capability, activities, and performance augment as Figure 1 illustrates.

METHODOLOGY

The research process to prepare this chapter went through steps that Figure 2 exhibits as follows:

1. The authors familiarized themselves with the data through review, reading, listening, and so forth.
2. Next, they organized and indexed the data for easy retrieval and identification.
3. They coded the data by clustering the data into groups and then identified themes.
4. They explored relationships between codes. The researchers then refined the themes and groupings.
5. Next, the authors developed a conceptual model (Figure 9) and incorporated pre-existing knowledge.
6. The last step before writing the chapter was to test the conceptual model against data (Table 1) to assure the validity of the conceptual model.
7. They wrote the chapter, including excerpts from original data, such as quotes from interviews to substantiate our results and findings.

Figure 2. Research methodology

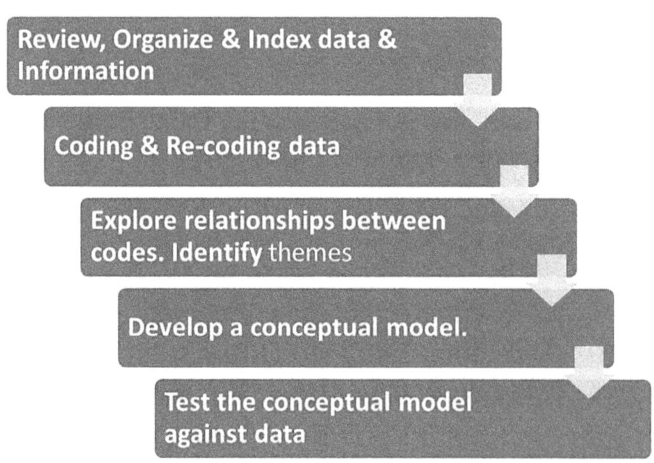

The chapter aims to explore the attributes of innovative cities and municipalities. Innovation is a process through which a society transforms itself into the desired state. For this reason, the research methodology the authors applied in this research is qualitative. The researchers collected data through reviewing and analyzing official reports, published papers (Al-Ansari, Altalib, & Sardoh, 2013; Al-Ansari, Pervan, & Xu, 2013; Bagwell, 2008; Elsholkamy, 2017; Elsholkamy & Fischbach, 2018), and articles from newspapers containing official declarations and opinions.

To augment and complement data collected from the document review, the authors designed a survey consisting of open-ended questions and questions with responses ranked on a Likert scale. They employed a specialized Internet survey website to collect responses via emails, Facebook groups, and similar tools. The researchers received 45 responses. They did not ask questions to which the respondent might feel uncomfortable (e.g., gender, age, and nationality). They were looking for their impressions and opinions while assuring their anonymity.

The authors used qualitative data software to analyze the data. They then compared data they gathered by the survey to those they collected from the documents as well as their notes and themes, such as the number of times an issue emerged, details and specificity, emotions (e.g., passion and happiness), and, lastly, how many different people mentioned something. The researchers used the software to generate word clouds that highlight issues the respondents raised, and a network to graphically represent the conceptual model depicting the innovative approach that DM employed to transform the city into a global, well connected, and smart city.

RESULTS

Efforts of the Government

DM has taken actions to transform the city in a short period into a global city that others envy and try to emulate. Dubai city embarked on a set of initiatives as articulated in Dubai 2020 and World Expo 2020, aiming to brand and transform the economy by enacting incentives for investment in services, ICT infrastructure, and urban development. This knowledge-based model is aligned with global best practices and empowered by smart infrastructure, legal reform, financial incentives, and innovations in service delivery in both the public and private sector (Acuto, 2010a; Hvidt, 2012, 2015). These innovative activities provide the conditions and capabilities to impact innovation performance. These initiatives are within an overarching framework for the UAE strategy. UAE Vision 2021 articulates a set of innovative urban technologies including the use of renewable energy, e-business, and digital transformations across all sectors, with an emphasis on health, education, sustainability, and urban planning. Also, Dubai invested in enhancing the enabling environment and conditions for public sector innovation. These include ICT infrastructure, accelerators programs, and technology parks. DM aimed to be among the top 20 countries of the UN Global Innovation Index and the annual revenue growth of 16%.

Dubai defined a set of metrics to measure levels of public sector innovation and to ensure a smart system of governance. Besides the financial measure, other measures include quality of public services, an index for mobile-based service adoption, the administrative efficiency index, and the rule of law index (Acuto, 2010b). Dubai started using the "Happiness Index" to measure residents and visitors' satisfaction. In essence, innovation in the public sector is receiving special attention in DM, as it hosts a global summit on public sector innovation to celebrate best practices worldwide. Dubai city is keen to empower and enables citizens to interact and contribute new ideas for efficient and quality service. The vision of DM is to develop an iconic global model of a smart city coupled with personalized e-services and open and user innovation to address the challenges of urbanization.

Customers' Opinion

According to TripAdvisor, a website dedicated to tourists, Dubai succeeded in blending contemporary culture with history, adventure with world-class shopping, and entertainment. The site recommends the Dubai Opera, Burj Khalifa, and spending an afternoon along Dubai Creek (i.e., Khor), discovering the gold, textile, and spice markets. Adventurous travelers can ride a hot air balloon that cruises over the desert dunes, ascent aboard a high-speed ride at Ilyas and Mustafa Galadari (IMG) Worlds of Adventure or skydive over the Palm Jumeirah.

An estimated 98% of the responses mentioned that tourism, conferences, and work are among the top reasons for visiting and living in Dubai, as Figure 3 indicates. Other reasons include visiting family, being born and being educated in Dubai.

Figure 3. Top reasons for visiting/living in Dubai

Shopping and urban attractions topped the list of the most favorite experience in Dubai (Figure 4). Respondents listed several iconic buildings, such as Burj Khalifa, the Metro, and Palm Jumeirah, as their most favorite experience. Global Village Dubai was one of the attractions that respondents mentioned as a reason for the favorite experience. It combines cultures from 90 countries in one place. Khor Dubai is a seawater inlet that the United Nations Educational, Scientific and Cultural Organization World Heritage Center added to its tentative list. It is another attraction that Dubai visitors and residents listed as a favorite experience.

Traffic jam, congestion, visa, and the weather were among the least favorite experiences in Dubai (Figure 5). One of the respondents wrote that traffic jam spoiled his/her enjoyment, where the response was "Nightmare traffic." Another respondent wrote that it is not easy to go from one place to another. According to a cab driver, he lives in Dubai with others in an apartment. With the rent he pays, he can get a studio in Sharjah, but he prefers living in Dubai to gain two extra hours of sleep that, otherwise, he would spend on the road commuting between Dubai and Sharjah, every day.

Traffic congestion is among the reasons for the degraded air quality of Dubai. One of the measures of air quality is the suspended particulate matter (PM) that consists of ashes, soot, aerosols and the like. PM produces health

Figure 4. Most favorite experiences

Figure 5. Least favorite experiences

hazards causing premature mortality and rates of morbidity. Burning fossil fuel is among the sources of PM. In 2013 and 2014, PM10 concentration reached 68.06 and 84.72 μg/m^3, respectively. The maximum World Health Organization acceptable level of PM10 is 50 μg/m^3. PM2.5 concentrations were 40.97 and 56.60 μg/m³ in 2013 and 2014, respectively. The maximum World Health Organization acceptable level of PM2.5 is 35 μg/m³ (World Council on Data City, n.d.).

The weather is the reason for the visitors' and residents' complaints. A respondent complained about the weather where s/he wrote: "Don't go during summer." Another commented on his/her least favorable experience as "hot." A respondent wrote that being in-door limited to his/her enjoyable time.

Foreigners working in Dubai have their worries regarding their status. A respondent who lives in Dubai wrote: "Renewing my employment visa was and still my nightmare. It is a process that makes me nervous." Another respondent's concern was deporting his/her spouse. Although Dubai invested in improving the performance of public institutions to ease businesses and free trade, it seems there is still a need to address issues about labor markets.

It seems that visitors are pleased with Dubai and are willing to visit/work in Dubai in the future. Approximately 60% of the respondents answered "Most likely I will return to Dubai," as Figure 6 indicates, which explains why the majority of the interviewees will recommend Dubai as a desirable tourist destination (Figure 7), thus reflecting the success of the adopted business model.

The majority of interviewees recommend Dubai for business (Figure 8), which is in line with the results that various institutions, such as Jones Lang LaSalle Incorporated, indicated. The efforts of the DM are among the reasons for the success of Dubai as a global city. Connectivity, job opportunities,

Figure 6. Likelihood of returning to Dubai

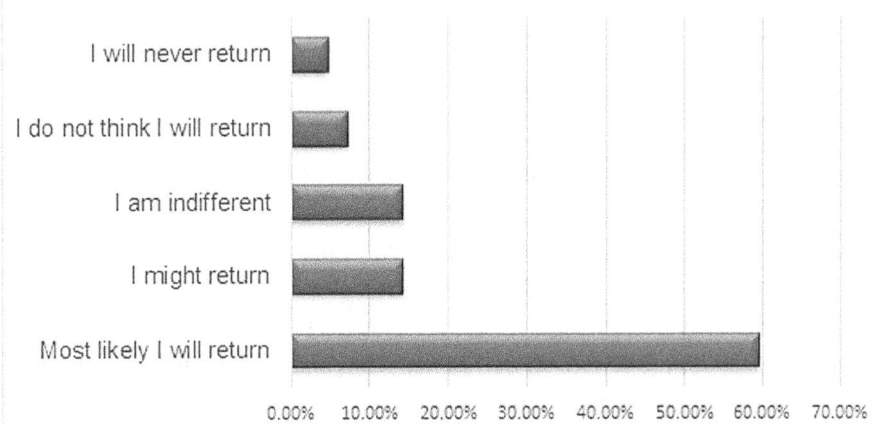

Figure 7. Most likely will recommend Dubai as a tourist destination

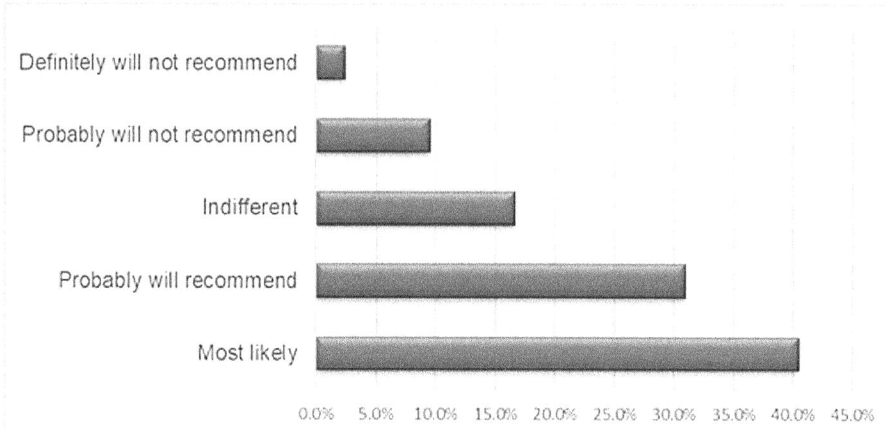

Figure 8. Likelihood of recommending Dubai for business

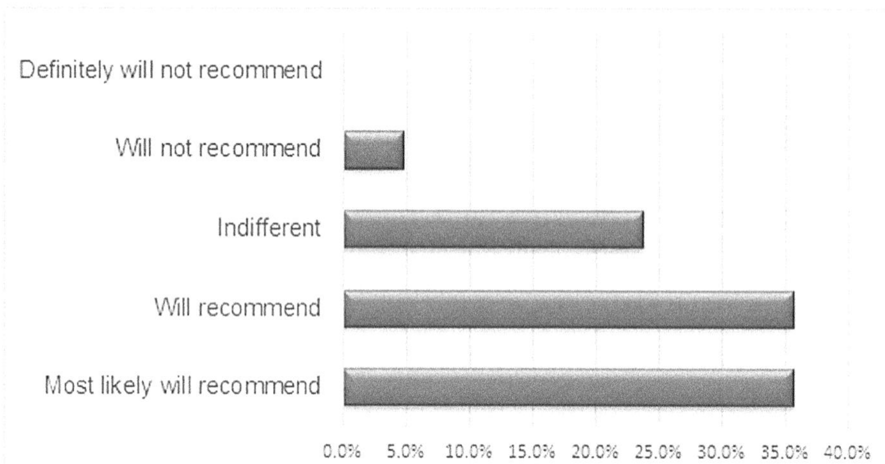

and business environment are among the key enablers that DM enjoys, thus transforming Dubai into a world tourist attraction and business hub.

The above assessments do not suggest that Dubai is the utopia that urban planners and philosophers seek. Being a tourist attraction and a city that serves the global economy is incomplete, as one of the respondents mentioned that the city is fake. Many take Dubai as it is, with its positive and negative aspects. Dubai is one of the facets of the latest stage of capitalism—cognitive-cultural capitalism, where the high-technology industry, corporate and financial services, personal amenities, media, and cultural industries are the leading sectors.

The culture of consumption and lifestyle can explain the mixed feelings and affection to Dubai. "Living in Dubai is like living in a blending mixer," said a resident of Dubai. The statement indicates contradictions. Another respondent summed the discussion as her relationship with the city seems complicated. She wrote:

My relationship with Dubai is bizarre and complicated. Hating Dubai is impossible; loving it is difficult. Although I have 100 reasons to hate Dubai, I can love the city for a million reasons. A consumer society up to the core, and socially segregated to the bottom. However, the city has developed some of the places where everyone melts on one land. Dubai is the forbidden love; it is everything except being humane. Dubai and I have a dreadful crack. Nevertheless, a thousand greetings to Dubai.

Three sets of themes explain the success of Dubai: 1) Innovation capability; 2) innovation activities; 3) innovation impacts. Each element includes sub-themes that work toward achieving the aim of DM, as Figure 9 indicates.

It is evident that innovation capability has the highest frequency (Table 1). This result indicates that a system thinking and sound holistic strategy and vision are the driving forces for an innovative municipality, indicating political determination and commitment. They are imperative to invest in infrastructure, systems of innovation, and technology adoption and diffusion.

The second highest attribute is the innovation activity, which entails the provision for a space for ideation, incubation, and implementation of ideas. It entails attention to human capital, governance, and talent management. Innovation activity includes selecting, assessing and implementing new ideas, indicating that DM is in line with a paradigm of open and user innovation through exhibitions, conferences, and awards to be part of a smart city movement that digital transformation is leading to new models of public sector innovations.

The improved performance has no occurrences in Table 1, probably because DM planned many initiatives beyond 2020, and the evidence of improved performance at many levels is limited. Nevertheless, the survey reveals a strong association between the efforts of DM and the users' satisfaction. However, the mindset and culture of urban innovation is evident in the branding and

Table 1. Co-occurrence table indicating relationships between different themes and codes

	Innovation Capability						Innovation Activity								Innovation Impact							
	Leadership & Culture		Management of Innovation		Organizational Enablers of Innovation		Assessing New Ideas		Diffusing What Works		Selecting & Developing Ideas		Implementing New Ideas		Improved Context		Improved Efficiency		Improved Organizational Performance		Improved Service Evaluation	
	Count	Coefficient	Count	Coefficient	Count	Coefficient	Count	Coefficient	Count	Coefficient	Count	Coefficient	Count	Coefficient	Count	Coefficient	Count	Coefficient	Count	Coefficient	Count	Coefficient
Leadership & Culture	0	0.00																				
Management of Innovation	5	0.38	0	0.00																		
Organizational Enablers of Innovation	6	0.46	4	0.31	0	0.00																
Assessing New Ideas	3	0.19	4	0.31	2	0.13	0	0.00														
Diffusing What Works	2	0.13	3	0.23	1	0.06	5	0.42	0	0.00												
Selecting & Developing Ideas	3	0.17	6	0.46	2	0.11	8	0.67	6	0.46	0	0.00										
Implementing New Ideas	3	0.14	5	0.29	2	0.10	8	0.53	7	0.47	10	0.67	0	0.00								
Improved Context	1	0.04	2	0.08	0	0.00	1	0.04	1	0.04	3	0.12	2	0.07	0	0.0						
Improved Efficiency	0	0.00	1	0.08	0	0.00	1	0.07	1	0.08	2	0.13	1	0.05	4	0.2	0	0.00				
Improved Organizational Performance	0	0.00	0	0.00	0	0.00	1	0.08	1	0.09	1	0.07	1	0.06	2	0.1	3	0.43	0	0.00		
Improved Service Evaluation	0	0.00	1	0.05	0	0.00	1	0.05	1	0.05	2	0.09	1	0.04	7	0.28	3	0.18	2	0.13	0	0.00

marketing of innovation through media, e-services, and celebrating diversity. Dubai combines the duality of both creative and innovative city, as the purposeful fusion of ideas and actions manifests.

DISCUSSION

Innovation Capability

Atop the factors that boosted the innovation capability of DM are leadership and culture. These two elements are the *raison d'être* for public innovation. DM is not making small plans. The administrators at DM are thinking big and thinking into the future. The goal of DM administrators is to transform their city into the world's most innovative city. In 2017, J. Chambers, the executive chairman of Cisco Systems, described Dubai as one of the most technologically advanced cities in the world. He then said that Dubai had placed itself firmly at the forefront of smart cities worldwide by merging the vision of its leadership with a unified strategy (Dubai Corporation of Tourism & Commerce Marketing, 2018).

Both leadership and culture are instrumental in encouraging frontline and managerial staff to present their ideas and proposals that help in attaining the objectives of the DM Strategic Plan and the directives of the Government of Dubai, as Director-General of DM expressed. He then continued "DM is keen to motivate its employees and teams to participate and win international awards, which reflects the efforts of the municipality and its integrated working systems in pioneering fields" (Admin, 2018).

In Dubai, innovative initiatives start with generating ideas. Dubai government and municipality employ various means for generating ideas, including, but not limited to, Internet-based research guidance, best practice information from similar organizations, and consultants. Also, Dubai generates ideas from conferences, trade fair, and exhibitions. For example, in 2018, Dubai hosted the Digital Innovation Summit for two days, where over 300 representatives from schools across the globe listened to 72 speakers who offered exceptional specialized development opportunities and a chance to present best practices ("Digital Innovation Summit I JESS Dubai," n.d.). In 2019, Dubai is organizing the Global Innovation Summit for top entrepreneurs and senior executives in charge of innovation and new product development across businesses and sectors within their business. Also in 2019, Dubai will

Figure 9. DM innovation model

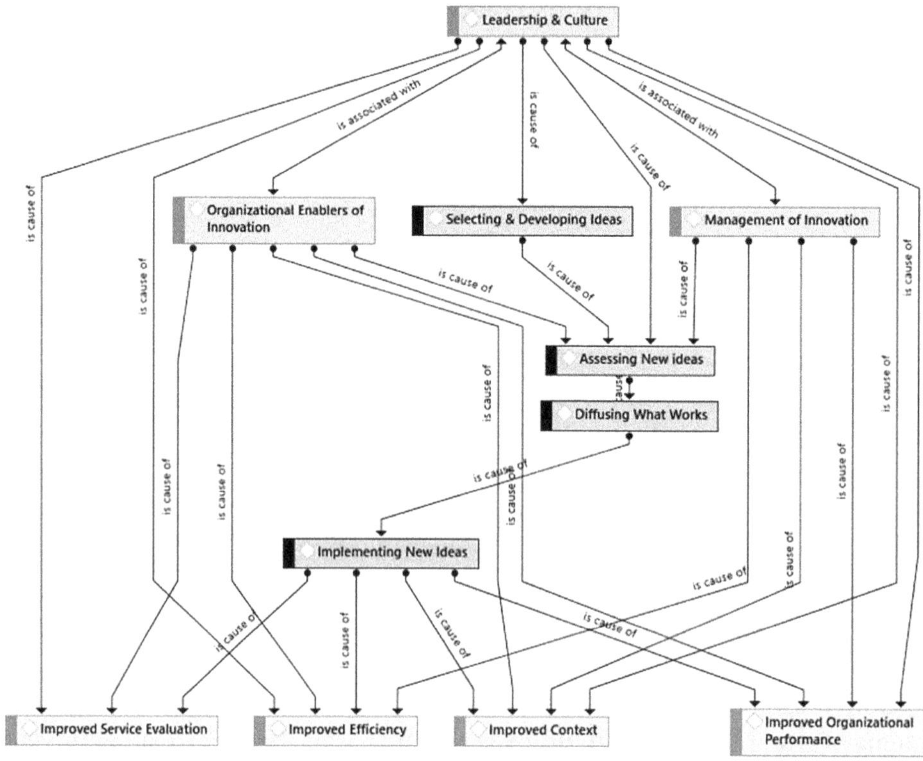

host Innovation Arabia 21, which intends to support Arab countries to build smart communities (Hamdan bin Mohammed Smart University (HBMSU), 2018). Over and above, Dubai established innovation labs and excellence awards to foster innovation in public sectors and other sectors of the economy, such as small and medium enterprises and startups.

Innovation Activities

In 2016, the World Economic Forum (WEF) named the UAE government as the second most tech-savvy government. Activities such as digitizing government services, introducing e-voting, and launching the re-chargeable e-dirham card are among the reasons for WEF's decision. The WEF acknowledgment is the outcome of the digital transformation initiative.

DM in collaboration with other institutions, such as the Dubai Electricity and Water Authority, embarked on executing innovative green technologies that help reduce carbon emissions. These initiatives can contribute to achieving the UAE's nationally determined contributions to the Paris Agreement under the United Nations Framework Convention on Climate Change.

The government did not just act as a leader that nurtured the culture of innovation but extended its role to be an enabler. The government established the Dubai Creative Clusters Authority (DCCA) to advance creative clusters in Dubai's education. The Advancement of Creative Clusters in Dubai's Education Sector: A Public-Private Partnership is a governmental initiative that allows institutions under the authority of DCCA to access a group of university students who present themselves as interns/part-time employees, looking for opportunities to gain practical experience in their fields of study, and increase their chances in the job market. DCCA is a government regulatory body. Its mandate is to assure that all involved parties (i.e., the university and the private sector employer) adhere to their obligations and protect their rights (Elsholkamy, 2018)). Thus, the government is enabling universities to succeed in accomplishing their mission, the private sector companies to attain their interests, and, most importantly, train students and developing their skills needed to meet real-life cases and problems.

Innovation Impacts

The outcome of innovation initiatives is evident in improved services, organization performance, efficiency, and enhanced context. The innovation activities the authors mentioned earlier have improved the working environment within DM. Their impacts extended to residents and visitors. With continued developments, issues and concerns, such a renewing the visa of a worker, will not continue to be an issue that affects his/her productivity.

The innovation impacts extend to include facilities and attractions to transformed Dubai to be one of the world's major attractions. These facilities and services are not just for the tourists, but also the residents, thus increasing their satisfaction to the extent that one of the interviewees mentioned "quality of life" as a positive aspect of living in Dubai.

To complete the above discussion, it is evident that innovative municipalities rise from visionary leadership, open-mid set and culture, and sound infrastructure that depends on technological advances. These parameters are critical components of the innovation system. The DM experience is a model for other cities of the GCC to emulate.

CONCLUSION

It is apparent that Dubai has succeeded to position itself among world cities through political will and determination, and meticulous planning. Adopting a creative and innovative approach to develop and select ideas is among the reasons for the success that Dubai enjoys today which is underpinned by enablers, processes, and innovation performance, Figure 10. Innovative human capital, organizations, and carefully elaborated and implemented plans are all aligned with the innovation system of the UAE.

DM enjoyed the support of the leaders of the Emirate and nurtured a culture conducive to innovation by enabling staff the opportunity to come up with ideas, hosting conferences, exhibitions and trade fairs, and exchanging ideas with others. DM successfully developed its innovation capability to implement innovation activities, which paid off, thus enjoying innovation outcomes. System thinking is the path that DM followed to foster innovative schemes rooted in a bold strategy, leadership and culture, and infrastructure, Figure 11. The strategy focuses on service innovation and harnessing global networks and alliances. Learning communities and knowledge management

Figure 10. Determinants for innovation and knowledge management in a, innovative municipality

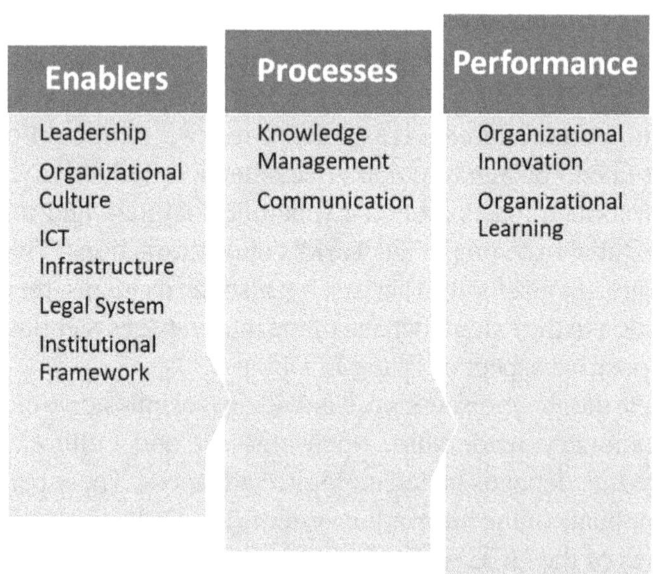

Figure 11. Enablers and factors for innovation systems in a smart municipality

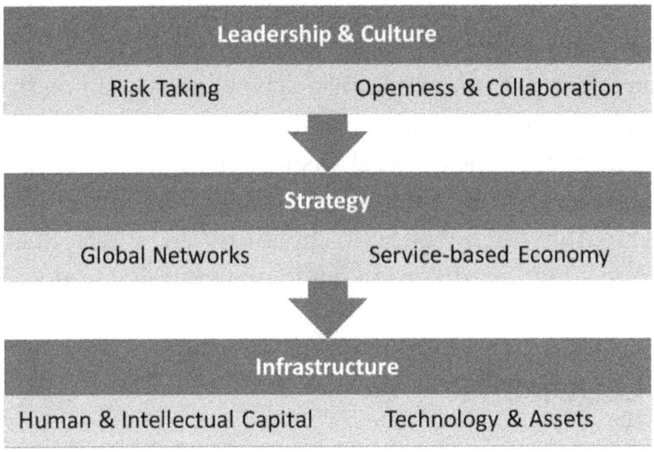

processes characterize the culture of DM. Also, investing in smart systems and infrastructure including airports and transport support innovations in ICT for green technologies and e-governance -- these are the cornerstones for an innovative city and municipality.

However, Dubai still has to address issues and problems. For example, traffic congestion is an issue that has impacts on air quality. The introduction of mass transit, such as the Metro, can help. It is certain that Dubai will solve these issues so long DM is adopting innovative solutions and investing in global innovation networks and strengthening the ecosystem of innovation.

Eliciting severl lessons from DM experience is possible. Without leadership and culture for innovation, there is a limited possibility that cities of GCC we be able to face critical challenges and be both sustainable and resilient. When city administrators have the vision, political will and determination, they will be able to transform their metropolises into smart cities. Technology adoption and fusion of ICT are crucial for cities of GCC to shift gears toward a post-oil era.

The center of gravity is moving toward to the Gulf Cooperation Council (GCC) countries in terms of technology and innovation which is a testimony of the sound incentives in terms of systems and processes to foster a sustainable and a resilient city. However, DM is to be prepared to have absorptive capacity in its innovation system to adapt to the 4th industrial revolution and the emerging disruptive technologies in public sector and health, education, and finance like fintech, blockchain, driverless cars, and renewable energy.

Future research agenda in the GCC context has to interrogate the following:

1. Revisit Science, Technology and Innovation (STI) agenda and align Research and Development (R&D) priorities to address local needs, particularly in water, food security, and green energy.
2. Examine the impact of technology-driven innovation on the social and ecological sustainability of the Gulf region
3. Assess the effectiveness R&D institutions and the efficiency of the National Innovation Systems within the region.
4. Investigate the organizational capability of R&D institutions for technology adoption, diffusion, and commercialization.

ACKNOWLEDGMENT

The authors would like to thank all those who participated in the survey. Special thanks to Ghada Yassine, Assistant Professor, Department of Architectural Engineering, Menofia University, Egypt, for reviewing earlier versions of the chapter.

REFERENCES

Acuto, M. (2010). High-rise Dubai urban entrepreneurialism and the technology of symbolic power. *Cities (London, England)*, *27*(4), 272–284. doi:10.1016/j.cities.2010.01.003

Admin. (2018). *'Dubai Municipality innovations' win 8 local, international awards*. Retrieved September 2, 2018, from http://emirates-business.ae/dubai-municipality-innovations-win-8-local-international-awards/

Al-Ansari, Y., Altalib, M., & Sardoh, M. (2013). Technology Orientation, Innovation and Business Performance: A Study of Dubai SMEs. *The International Technology Management Review*, *3*(1), 1. doi:10.2991/itmr.2013.3.1.1

Al-Ansari, Y., Pervan, S., & Xu, J. (2013). Innovation and business performance of SMEs: The case of Dubai. *Education, Business and Society*, *6*(3/4), 162–180. doi:10.1108/EBS-04-2013-0012

Al-Jayyousi, O. (2004). Knowledge creation in the water sector: Towards a learning water organization. *International Journal of Water Resources Development*, *20*(2), 165–175. doi:10.1080/0790062042000206147

Al-Jayyousi, O. (2017). *Integral Innovation*. New York, NY: Taylor and Francis. doi:10.4324/9781315588957

Anon. (n.d.). *Dubai Population*. Retrieved December 7, 2018, from http://worldpopulationreview.com/world-cities/dubai/

Anttiroiko, A. V. (2016). *City-as-a-platform: The rise of participatory innovation platforms in Finnish cities* (Vol. 8). Sustainability. doi:10.3390u8090922

Arundel, A., & Hollanders, H. (2011, October). A Taxonomy of innovation: How Do Public Sector Agencies Innovate. *European Commission*, 1–38.

Asheim, B., Isaksen, A., Moodysson, J., & Sotarauta, M. (2011, January). Knowledge bases, modes of innovation and regional innovation policy. *Beyond Territory: Dynamic Geographies of Innovation and Knowledge Creation*, 227–249.

Bagwell, S. (2008). Creative clusters and city growth. *Creative Industries Journal*, *1*(1), 31–46. doi:10.1386/cij.1.1.31_1

Bankins, S., Denness, B., Kriz, A., & Molloy, C. (2017). Innovation Agents in the Public Sector: Applying Champion and Promotor Theory to Explore Innovation in the Australian Public Service. *Australian Journal of Public Administration*, *76*(1), 122–137. doi:10.1111/1467-8500.12198

Borins, S. (2001). Encouraging innovation in the public sector. *Journal of Intellectual Capital*, *2*(3), 310–319. doi:10.1108/14691930110400128

Borins, S. (2016). *The Challenge of Innovating*. In Government Innovation Series Sandford Borins Professor of Public Management University of Toronto.

Caulier-grice, J., & Mulgan, G. (2010). And Grow Social Innovation the Open Book of Social Innovation. *Young*, *30*(8), 224. doi:10.1057/9780230595040

Cucciniello, M., & Nasi, G. (2014a). Evaluation of the Impacts of Innovation in the Health Care Sector: A comparative analysis. *Public Management Review*, *16*(1), 90–116. doi:10.1080/14719037.2013.798026

Cucciniello, M., & Nasi, G. (2014b). Transparency for Trust in Government: How Effective is Formal Transparency? *International Journal of Public Administration*, *37*(13), 911–921. doi:10.1080/01900692.2014.949754 de Vries, H., Bekkers, V., & Tummers, L. (2016). Innovation in the Public Sector: A Systematic Review and Future Research Agenda. *Public Administration, 94*(1), 146–166. doi:10.1111/padm.12209

Digital Innovation Summit | JESS Dubai. (n.d.). Retrieved December 7, 2018, from https://www.jess.sch.ae/jess-conferences/digital-innovation-summit

Dubai Corporation of Tourism & Commerce Marketing. (2018). *Dubai Future City - Aiming to be the most innovative city by 2021*. Retrieved November 30, 2018, from https://www.visitdubai.com/en/business-in-dubai/why-dubai/news-and-insights/dubai-dreams-big-innovation-future

El-Kholei, A. O. A. O. (2018). Development of Planning Theory and Profession: Implications for Planning Education and Practice. *Archnet-IJAR, 12*(1), 182–208. doi:10.26687/archnet-ijar.v12i1.1503

Elsholkamy, M. M. (2017). The Advancement of Creative Clusters in Dubai ' s Education Sector : A Public-Private Partnership towards Development. In *Global Applied Business Research Conference* (pp. 1–20). London, UK: University of London.

Elsholkamy, M. M. (2018). Creative Clusters in Dubai's Education sector: A Public-Private Partnership towards development. *UAE Public Policy Forum*. Retrieved from https://www.researchgate.net/publication/327720017_Creative_Clusters_in_Dubai's_Education_sector_A_Public-Private_Partnership_towards_development

Elsholkamy, M. M., & Fischbach, T. (2018). *Case Study The Dubai Municipality and the Excellence Award : Innovation in Procurement Management*. Dubai, UAE: Academic Press.

Geels, F. W., & Schot, J. (2007). Typology of sociotechnical transition pathways. *Research Policy, 36*(3), 399–417. doi:10.1016/j.respol.2007.01.003

Godin, B. (2007). National Innovation System: The System Approach in Historical Perspective. *Science Technology And Human Values, 34*(476–501), 1–37. doi:10.1177/0306312706075338

Gow, J. I. (2014). Public sector innovation theory revisited. *The Innovation Journal, 19*(2), 1–23. doi:10.1016/j.brat.2011.03.004

Hall, P. (2001). Cities in Civilization: Culture, Innovation, and Urban Order. *Plan Canada*, *41*(3), 13–15. doi:10.1093/ehr/116.465.155

Hamdan bin Mohammed Smart University (HBMSU). (2018). *Innovation Arabia 2019*. Retrieved December 7, 2018, from http://www.innovationarabia.ae/

Hughes, A., Moore, K., & Kataria, N. (2011). Innovation in Public Sector Organisations, A pilot survey for measuring innovation across the public sector. *NESTA Index Report*. Retrieved from https://www.nesta.org.uk/documents/108/innovation_in_public_sector_orgs.pdf

Hvidt, M. (2012). Planning for Development in the GCC States:A Content Analysis of Current Development Plans. *Journal of Arabian Studies*, *2*(2), 189–207. doi:10.1080/21534764.2012.735454

Hvidt, M. (2015). Transformation of the Arab Gulf Economies into Knowledge Economies : Motivational Issues Related to the Tertiary Education Sector. Doha, Qatar: Academic Press.

Inkinen, T. (2015). Reflections on the innovative city: Examining three innovative locations in a knowledge bases framework. *Journal of Open Innovation: Technology, Market, and Complexity*, *1*(1), 8. doi:10.118640852-015-0009-5

Kleiman, N., & Hillard, T. (2016). *Innovation and the City*. Academic Press.

Kuhlmann, S., & Rip, A. (2014). The challenge of addressing Grand Challenges: a think piece on how innovation can be driven towards the "Grand Challenges" as defined under the prospective European Union Framework Programme Horizon 2020.(No. 11). Enschede, The Netherlands: Academic Press.

Moussa, M., McMurray, A., & Muenjohn, N. (2018). A Conceptual Framework of the Factors Influencing Innovation in Public Sector Organizations. *Journal of Developing Areas*, *52*(3), 231–240. doi:10.1353/jda.2018.0048

Population. (2018). *Population of Dubai: Current Population Statistics*. Retrieved December 7, 2018, from https://www.dubai-online.com/essential/population/

Randhawa, K., & Scerri, M. (2015). Service Innovation: A Review of the Literature. In R. Agarwal, W. Selen, G. Roos, & R. Green (Eds.), *The Handbook of Service Innovation* (pp. 27–51). London, UK: Springer-Verlag; doi:10.1007/978-1-4471-6590-3_2

Sanders Jones, J. L., & Linderman, K. (2014). Process Management, Innovation, and Efficiency Performance: The Moderating Effect of Competitive Intensity. *Business Process Management and Innovation*, *20*(2), 335–358. doi:10.1108/BPMJ-03-2013-0026

Simmie, J. (2005). Innovation and space: A critical review of the literature. *Regional Studies*, *39*(6), 789–804. doi:10.1080/00343400500213671

Swan, M. (2014, June 30). Dubai No 2 most-expensive city to stay in, report says. *The National*. Retrieved from https://www.thenational.ae/business/travel-and-tourism/dubai-no-2-most-expensive-city-to-stay-in-report-says-1.605141

Teofilovic, N. (2002). The Reality Of Innovation In Government. *The Innovation Journal*, *7*(3), 1–30.

Turnheim, B., & Geels, F. W. (2013). The destabilisation of existing regimes: Confronting a multi-dimensional framework with a case study of the British coal industry (1913–1967). *Research Policy*, *42*(10), 1749–1767. doi:10.1016/j.respol.2013.04.009

Visvizi, A., Lytras, M. D., Damiani, E., & Mathkour, H. (2018). Policymaking for smart cities: Innovation and social inclusive economic growth for sustainability. *Journal of Science and Technology Policy Management*, *9*(2), 126–133. doi:10.1108/JSTPM-07-2018-079

Walker, J. L., Price, D., Browning, R., Miller, W., Mohr, L., Friedman, R., … Walker, L. (2008). The Diffusion of Innovations among the American States. *The American Political Science Review, 63*(3), 880-899.

Yun, J. J., Jeong, E., & Yang, J. (2015). Open innovation of knowledge cities. *Journal of Open Innovation: Technology, Market, and Complexity*, *1*(1), 16. doi:10.118640852-015-0020-x

Chapter 3
A Bibliometric Perspective on Technology-Driven Innovation in the Gulf Cooperation Council (GCC) Countries in Relation to Its Transformative Impact on International Business

Raheem Sarwar
https://orcid.org/0000-0002-0640-807X
City University of Hong Kong, Hong Kong

Saira Hanif Soroya
University of the Punjab, Pakistan

Amina Muazzam
Lahore College for Women University, Pakistan

Fahad Sabah
Superior University, Pakistan

Sehrish Iqbal
Information Technology University, Pakistan

Saeed-Ul Hassan
Information Technology University, Pakistan

ABSTRACT

This chapter presents a novel scientific research landscape of the Gulf Cooperation Council (GCC) in order to access the research productivity, scholarly impact, and international collaborations across all GCC countries over the time period of 2008–2018, using the Scopus database. While we observe a significant increase in investing the resources in GCC world in

DOI: 10.4018/978-1-5225-9012-5.ch003

order to build research infrastructure to make shift from only oil producing countries to knowledge-based economies, not enough efforts have been done to measure the outcomes of these investments. The chapter explores different thematic areas that have evolved over a decade related to TDI, top scientists, and top institutions in GCC. Further, the study investigates international collaboration networks within or outside GCC to better understand the venues of knowledge exchange with GCC countries. It is expected that the findings of this case study would provide an insight to the research landscape of the GCC and useful information to the scientific community as well as to the technology and innovation policymakers.

INTRODUCTION

Technology-driven innovations and scientific research play an important role in the economic growth of a country by producing the cost effective solutions and quality improvement of their products. Thus, it is impossible for countries and societies to develop sustainable economies without taking initiatives towards the enhancement of technology-driven innovation and scientific research (Hassan, Sarwar, & Muazzam, 2016). Modern economies are established on the bases of technology-driven innovation and scientific research and it has been the key focus of modern nations. Thus, the socio-economic and environmental conditions of these modern nations have been improved. The policymakers and the governments of these nations have adopted countless approaches to upsurge the technology-driven innovations (Sarwar & Hassan, 2015).

The *Gulf Cooperation Council (GCC)* nations proposed a vision of transforming their petrochemical economies to sustainable and diversified economies with the help of technology-driven innovations and scientific research. GCC nations have significantly invested in fostering the scientific research and technology-driven innovations as these are two most effective factors that help to transform a country's economy into a knowledge-based economy. Specifically, GCC nations have established state-of-the-art research centers, universities, and knowledge parks. Besides that, GCC nations have established the campuses of prestigious international universities to bridge the gap (Wiseman & Anderson, 2012). Some of the initiates taken by GCC nations are illustrated in the following paragraphs of this section.

Arab Science and Technology Foundation (ASTF) was established by 2000 in Sharjah City and it is a non-profit, non-governmental and civil organization aims to support technology-driven innovation and scientific research to transform United Arab Emirate's economy into a knowledge-based economy. This organization has supported more than 700 researchers working in more than 140 different projects and it has established 30 companies focused on R&D (cf.: http://www.astf.net).

In 2007, the government of United Arab Emirates (UAE) established the worlds' largest Higher Education Free Zone known as *Dubai International Academic City (DIAC)*. DIAC education zone also aims at transforming the UAE economy into a knowledge-based economy by 2021. Currently, DIAC is hosting a community of 24,000 students from 145 nationalities. These students have access to over 400 different higher education programs. The government of UAE also proposed a vision 2021 which aims at enlarging the socio-economic developments. Specifically, it seeks to transform the UAE petrochemical economy into a knowledge-based economy. To fulfill the objectives of the vision 2021, the government of UAE has increased the Research and Development budget. Recently the government of UAE has announced $82 billion funding to facilitate the vision 2021. Besides that, several research centers have been established in the country, such as, Science & Innovation Park was established by United Arab Emirates University and the Khalifa Innovation Centre was established by Khalifa University of Science & Technology. In addition to this, Abu Dhabi, the capital state of UAE is capitalizing on its energy resources to transform its economy to a sustainable economy. The Masdar Institute is at the core of this planning which is attracting powerful partner enterprises like Siemens and Credit Suisse and plans to attract 50, 000 peoples. In Abu Dhabi, the plan is to reduce the long-term dependency on hydrocarbons and maintain status in the energy sector (Adams et al. 2011).

To promote the technology-driven innovations, the kingdom of Saudi Arabia announced the national science technology and innovation plan (NSTIP) in 2007. The government of Saudi Arabia believes that the technology-driven innovation is the next economically effective force of the kingdom after the oil industry. The Secretary General of NSTIP, Dr. Alabdakadir said that "the NSTIP has its strategic goals and an ambitious long-term vision to transform the economy and society of the Kingdom into knowledge-based economy and society through a globally competitive national array". Since NSTIP is established, Saudi Arabia has marked tremendous growth in scientific publications. A part of NSTIP was to promote the public-private partnership aiming at innovation developments. For instance, King Abdullah University of Science and Technology (KAUST) is collaborating with Saudi Arabia Advanced Research Alliance (SAARI) to enhance the progress of innovation and economic development.

Moreover, the national strategy of the government of Saudi Arabia announced in 2012 aims to transform its economy into a knowledge-based economy. In addition to this, the government of Saudi Arabia released the vision 2030 which aims at completing 80 projects costing up to $3.7 million (Al-Yousif, 2008).

It can be observed that the government of Qatar has made a significant increase in investing the resources to build research infrastructure, and to make a shift from oil producing country to a knowledge-based economy. Specifically, the government of Qatar has improved the higher education system in the country and has strengthened the resources required to achieve its goals. Qatar Foundation is one of the main organizations providing funds to build research and Development (R&D) centers across the country. The prominent R&D centers build by Qatar foundation are: Qatar Computing Research Institute (QCRI), Qatar Environment and Energy Research Institute (QEERI) and Qatar Biomedical Research Institute (QBRI). Moreover, Qatar foundation supervises the Qatar Science & Technology Park (QSTP) and Sidra Medical and Research Center with a flagship of $7.9 billion. In addition to this, Hamad bin Khalifa University and Qatar universities are conducting useful research to transform Qatar into a knowledge-based economy. Similar to UAE, Saudi Arabia, and Qatar other GCC nations such as Jordan, Oman and Bahrain have also taken several initiatives to transform their economies into knowledge-based economies (Al-Yousif, 2008; Wiseman & Anderson, 2012).

This chapter presents a novel scientific research landscape of the Gulf Cooperation Council (GCC) to access the research productivity, scholarly impact and international collaborations across all GCC countries over the period of 2008–2018, using the Scopus database. GCC is an economic and political alliance of six middle eastern countries including United Arab Emirates, Saudi Arabia, Qatar, Oman, Jordan, and Bahrain. The GCC was established in May 1981, in Riyadh Saudi Arabia (Alajmi & Rorissa, 2018). While a significant increase in investing the resources in GCC world in order to build research infrastructure to make shit from only oil producing countries to knowledge-based economies can be observed. However, not enough efforts have been done to measure the outcomes of these investments. This chapter is an effort to fill the gap by providing an in-depth bibliometric perspective of scientific research in the areas related to Technology Driven Innovation (TDI) and its impact on international business using the scientific literature. This analysis explores different thematic areas that have evolved over a decade related to TDI, top scientists, and top institutions in GCC. Further, we investigate international collaboration networks with-in or outside GCC to better understand the venues of knowledge exchange with GCC countries. We believe the findings of this case study would provide an insight to the research landscape of the GCC and useful information to the scientific community as well as to the technology and innovation policy makers. The rest of the chapter is organized as follows. Section 2 reviews the literature. Whereas, section 3 presents the data collection and methodology. Section 4 contains the discussion on findings of this chapter, and finally, conclusions are drawn based on the findings of the study.

LITERATURE REVIEW

A large percentage of economic development is dependent on knowledge-driven innovation and technological transformation. No matter the section, innovation is linked to products or procedures, it is sudden or stepwise in nature, or it is technical or associated to business prototypes, it is an essential ingredient for advancement of a nation(Karkouti, 2016).

The ability of a society to innovate and withstand the growth depends upon the effectiveness of policies and the connections among institutions. The expressions "national innovation ecosystem" (NIE) and "national innovation system" (NIS) have become a characterization of the association between people, initiatives and institutions that lead the transfer of information & technology in a country (Nelson, 1993). Both of these terms contain components such as human resource, research & development, networks that assist in quick communication and collaboration between technical specialists, an atmosphere for business, a policy environment, and infrastructure. A softness in any of these component can create a noteworthy barrier to innovation.

Studies have shown that the innovation capability has backed up the economic growth of many countries in recent years (Fan, 2011). The improved innovation ability was principally the result of heavy investments in research & development (R&D). Governments have redefined their NIEs by connecting the science and business sectors and balancing the indigenous R&D efforts with technology import (Motohashi, 2005). In addition to this, other secters such as relationships in R&D objects and business plays a significant part to formulate make the national innovation ecosystem. Southeast Asian countries articulated their NIE structures to support developing economies to work on an industrial-development policy (Wong, 2011). Industrial-development policy in the countries, local technologies benefit from connections and technological collaborations, is evident in the freshly industrialized economies.

Most of the GCC countries represent a case in which capital formation does not enforce any constraint on development. The GCC countries have been motivated on technology transfer to recompense for a rareness of local technological capabilities. Three main elements are involved: manpower, disbursement of funds, and a well-conceived strategy for getting better out of scientific and technological activities. The five-year national development plans the Saudi government inaugurated in the mid-1970s (eight of them so far) have been an attempt to position the development regularity inside the country, holding its prominence as a up-to-date state without losing its headship position as the heart of the Islamic world.

Regardless of the substantial attainments in terms of producing an up-to-date economic infrastructure and refining their scholastic system (Karkouti, 2016), the research proposes that the superiority of education system in the GCC countries lacks the fundamentals that organize students for the industry. This is primarily due to the reason that the practical training and science & information technology programs are not stressed in their academic curricula (Bahgat, 1999; Wiseman & Anderson, 2012). Today, companies in Qatar require skillful experts in the fields of technology and industry in order to withstand the nation's development on the economic front (Al-Jaber & Dutta, 2008; Stasz et al., 2007).

In order to address these problems and to develop a knowledge-based society, the government of Qatar has created the Supreme Council of Information and Communication Technology (ictQatar) in 2004 for boosting the use of innovation and technology through generating a cutting-edge information and communication technology framework that encourages ICT in schools and improves human capital of Qatar (Nasser et al., 2011; Naseer & Romanowski, 2011). Al-Jaber and Dutta (2008) pointed out that ictQatar is the supervisor and supporter of the ICT segment and is responsible for creating an online knowledge-based community, founding the country as a technologically advanced nation, and improving the nation's technology literateness in order to overcome the country's shortage of professional workers.

DATA COLLECTION AND METHODOLOGY

As mentioned earlier in the introduction section, this chapter presents a novel scientific research landscape of the Gulf Cooperation Council (GCC) to access the research productivity, scholarly impact and international collaborations across all GCC countries over the time of 2008–2018, using the Scopus database. The following query was used to fetch data from Scopus.com dated: Sep 30, 2018: "*title-abs-key (technology) and title-abs-key (innovation) and (affilcountry(bahrain) or affilcountry(kuwait) or affilcountry(oman) or affilcountry(qatar) or affilcountry (saudiarabia) or affilcountry (united arab emirates))*".

In order to achieve the objectives of this study, we performed five different studies. The first study reports the publication productivity and research impact of GCC countries during January 1990 to September 2019. The second study reports the evolution of scientific literature in terms of publication count and key terms. Third study discusses the co-authorship network of GCC countries at author's level. These kinds of analysis could be found helpful to generate new and cherished information related to the planning, implementation, and monitoring of the research program. Forth study discusses the co-authorship network of GCC countries' scientific literature in terms of publication count. Fifth study discusses the citation networks at sources level of the publications of GCC countries from January 1990 to September 2018. Sixth and the last study discusses keywords co-occurrence analysis of the research publications of GCC countries from January 1990 to September 2018. Keywords analysis is especially suitable to get the overview of a field as they show the main theme of the publication

In terms of bibliometric indicators, we used publication volume, citation counts and number of collaborative publications to access the research productivity, scholarly impact and international collaborations across all GCC countries, respectively. While analyzing the intra- and inter collaboration network of GCC countries, we used the standard double counting approach which gives the equal publication credit to all collaborating countries. As for the visualization of results, the figures 1 and 2 are created using Tablea software and the rest of the figures are created using VOS Viewer software.

RESULTS AND DISCUSSIONS

Publication Productivity and Research Impact of GCC Countries

This section presents the publication productivity and research impact of GCC countries during January 1990 to September 2018. As can be seen from Figure 1(a) that United Arab Emirates has the highest number of publications among GCC countries. On the other hand, Saudi Arabia has a high number of citations among the GCC which shows that Saudi Arabia is more concerned with the quality and impact of its research as shown in Figure 1(b). Moreover, the publication and citation count of GCC nations shown in Figure 1 conforms to research expenditure on Research and Development (R&D) by these countries mentioned in the Introduction section. Overall results predict that Saudi Arabia has the necessary resources to invest in people, infrastructure and R&D. As Figure 1 depicts that in GCC only Saudi Arabia produced an adequate number of research publications with enhanced citations.

Figure 1a. Publication counts from January 1990 to September 2018

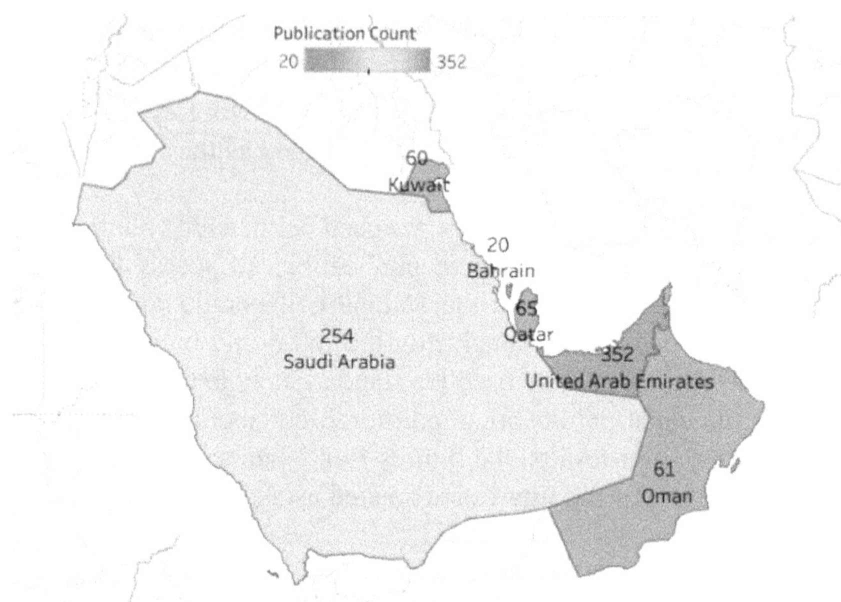

Figure 1b. Citation counts from January 1990 to September 2018

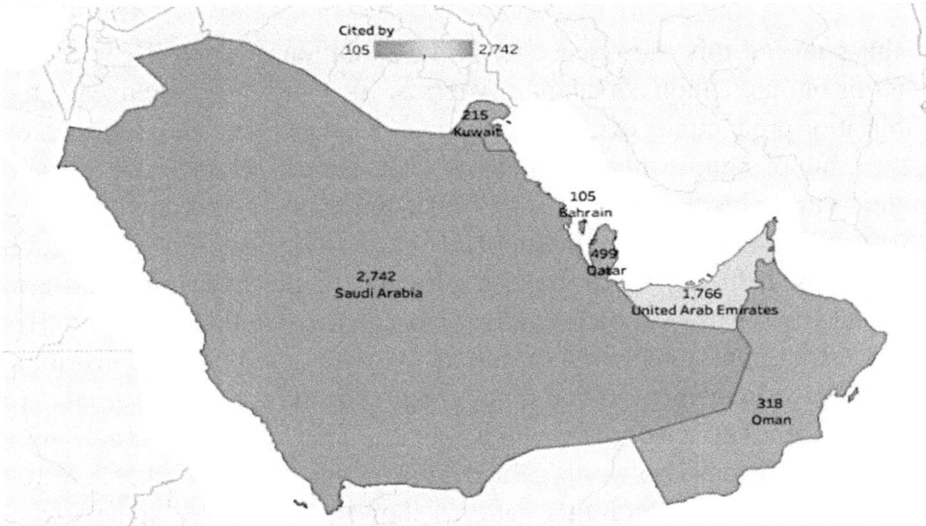

Table 1 shows the GCC countries publication counts and the citations received over the span of January 1990 to September 2018. It can be seen that Saudi Arabia has the second highest number of publications which received the highest citations as well as the highest average citation per paper. Whereas Qatar has 65 publications and 499 citations count which comes out to be the second highest average citation per paper. Bahrain, Oman, and United Arab Emirates have 20, 61 and 352 publications respectively and all of these three countries received almost equal average citations per paper. Then there is Kuwait which has 60 publications and it has lowest average citations per paper.

Table 1. GCC countries w.r.t. publication counts from January 1990 to September 2018

Country	Publication Count	Citation Count
Bahrain	20	105
Kuwait	60	215
Oman	61	318
Qatar	65	499
United Arab Emirates	352	1,766
Saudi Arabia	254	2,742

Evolution of Scientific Literature in GCC Nations

In this phase of this chapter, we study the evolution of scientific literature in terms of publication count and key terms. As depicted in Figure 2(a), the publication productivity of GCC nations have been increase after 2004, which is the result of significant investments of GCC nations on scientific research in the form of Saudi Arabia's vision of becoming an international leader in science and technology as well as council of Information and Communication Technology (ictQatar). As can be seen in Figure 2(b) that most important term in recent years is innovation which leads to a conclusion that GCC countries are investing in research and innovation. We also note that the E-learning, knowledge management, cloud computing, education, technology etc. are other frequent key term in the publications of GCC countries. The Figure 2(b) shows that these keywords gained more attention of researcher in recent years. The key terms evolution patterns showed that the researchers from GCC countries are trying to expand the scope of innovation and technology in research and it is becoming important in the research field.

Figure 2a. Evolution of scientific literature from January 1990 to September 2018

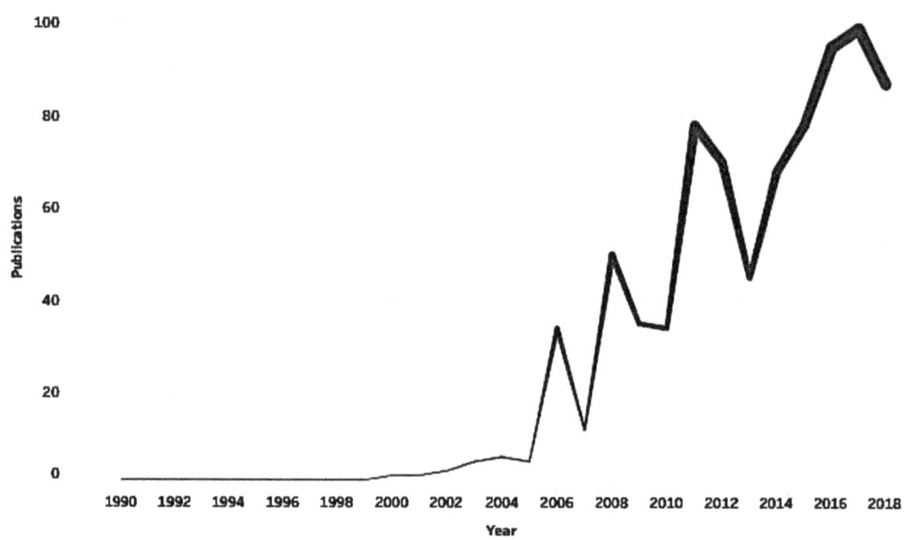

Figure 2b. Evolution of key terms from January 1990 to September 2018

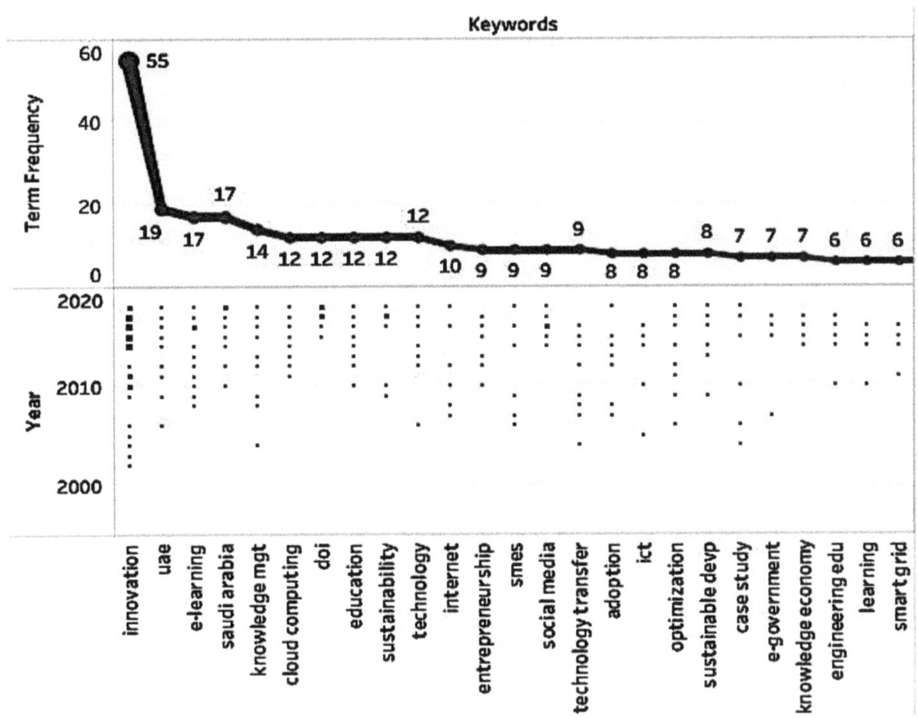

Co-Authorship Network From January 1990 to September 2018

This section of the chapter discusses the co-authorship network of GCC countries at author's level. These kinds of analysis could be found helpful to generate new and cherished information related to the planning, implementation, and monitoring of the research program. More practical role of the funding organizations regarding scientific capacity building objectives can be afforded and a more reliable engagement of institutions and authors from widespread regions based on innovative standards and parameters could be anchored in research. Collaboration between researchers is an important factor not only for the better research output but also for interacting with interrelated of one's discipline in this time of an innovative and knowledge-based society (Shin & Kim, 2018). Collaborations on international level had a great impact on research performance and they gain more benefit than local collaboration (Sabah et al., 2018).

We have drawn the co-authorship network of GCC countries from January 1990 to September, 2018, As can be seen in Figure 3 non-Arabic authors are well connected which means that most of the research conducted by GCC countries is being done with the international collaborations.

It has been observed that a fairly large number of highly cited researchers indicate some GCC countries' universities as a secondary affiliation address, thus enhancing these institutions up in international university rankings. The publications in which this happens are considered as internationally co-authored publications and can be expected to increase the GCC countries' ICAP as well.

Co-Authorship Network Among GCC and Other Countries From January 1990 to September 2018

In this phase of the chapter, we discuss the co-authorship network of GCC countries' scientific literature in terms of publication count. In Figure 4 the nodes represent the countries and the sizes of nodes show the extent of their collaborations whereas edges represent the number of collaborations between countries. It can be seen in Figure 4, that GCC countries mostly collaborates with countries other than GCC. As we observed earlier in Figure 3 that most of the authors are from countries other than GCC, it can also be seen in Figure

Figure 3. Co-authorship Network from January 1990 to September 2018

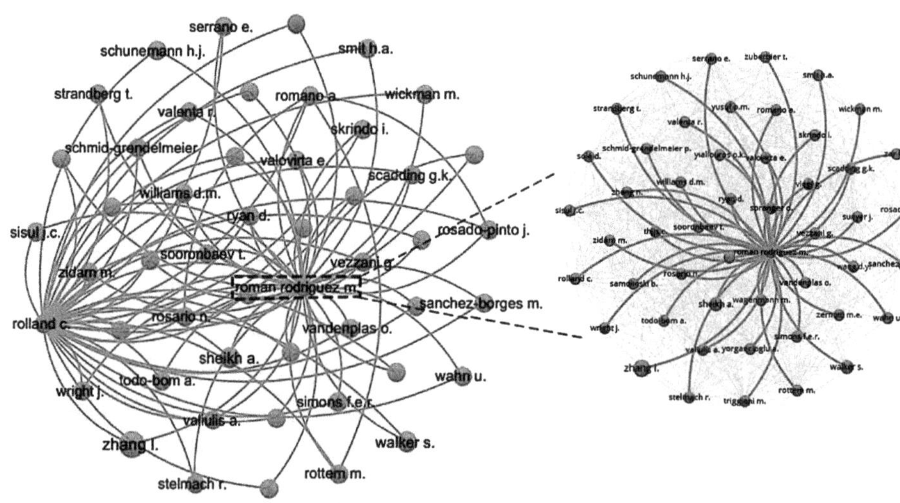

Figure 4. Co-authorship network among nations from January 1990 to September 2018

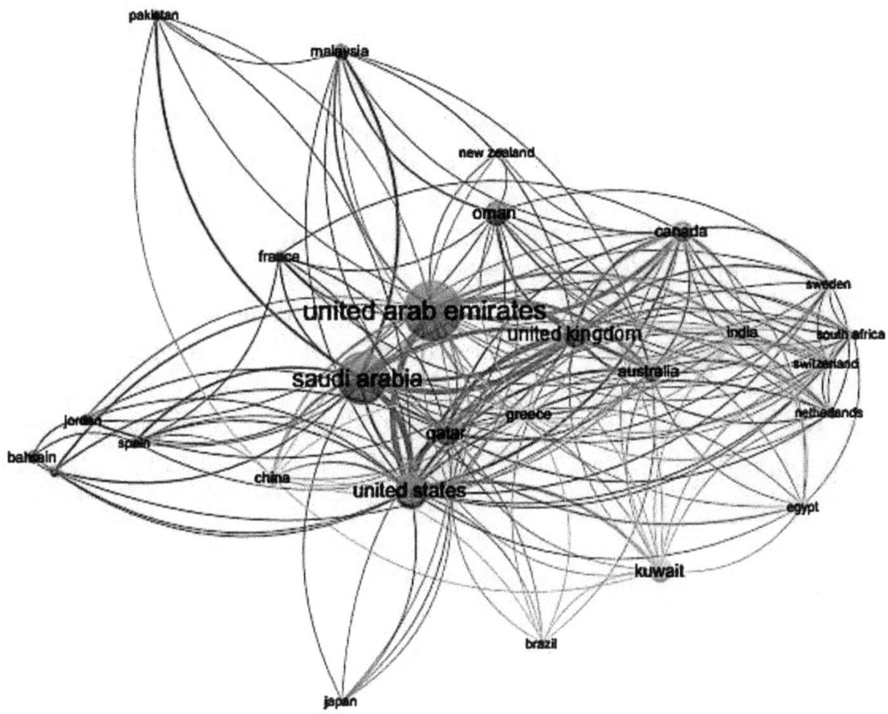

**For a more accurate representation see the electronic version.*

4 that the GCC countries mostly are more well connected with non-Arabic countries like; United States of America, China, France, and United Kingdom. Morover, Saudi Arabia also collaborates largely with Malaysia and Pakistan.

From these findings we can conclude that Saudi Arabia plays a dominating position of technological innovation aptitude in GCC, and USA, China, France, and United Kingdom are top countries of collaborative innovation with GCC. Based on our findings we observed that technological innovation competency of GCC countries has increased over time, but GCC countries genuinely depend on other advanced countries in technological innovation. The percentage of internationally co-authored publications (ICAP) of Saudi Arabia is quite high. The other two GCC countries namely; Qatar and UAE also showed an increase in their publication output, with ICAP percentages in 2015 of 87 and 68, respectively.

Citations Network Among Source Titles From January 1990 to September 2018

This part of the chapter discusses the citation networks at sources level of the publications of GCC countries from January 1990 to September 2018. Citation networks are visual maps in which the size of a node is proportional to the number of times a document, journal or conference has been cited by another document, journal or conference in addition to the VOSviewer calculation of total link strength. The total link strength is a way of measuring the strength of the relationship between nodes. Whereas nodes located near indicate more closely related to each other. The distance among the nodes shows the relativeness of the topics. In this network, closeness specifies related topics, whereas a larger distance specifies less associated fields of study. The visualization through figure 5 suggests that much of the work revolves around topics of innovation and technology, which will be examined more closely in the discussion later.

In figure 5, some source titles are more significant and have comparatively high number of citations like an international journal of innovation and proceedings of 28th international business information management had the

Figure 5. Citations network among source titles from January 1990 to September 2018

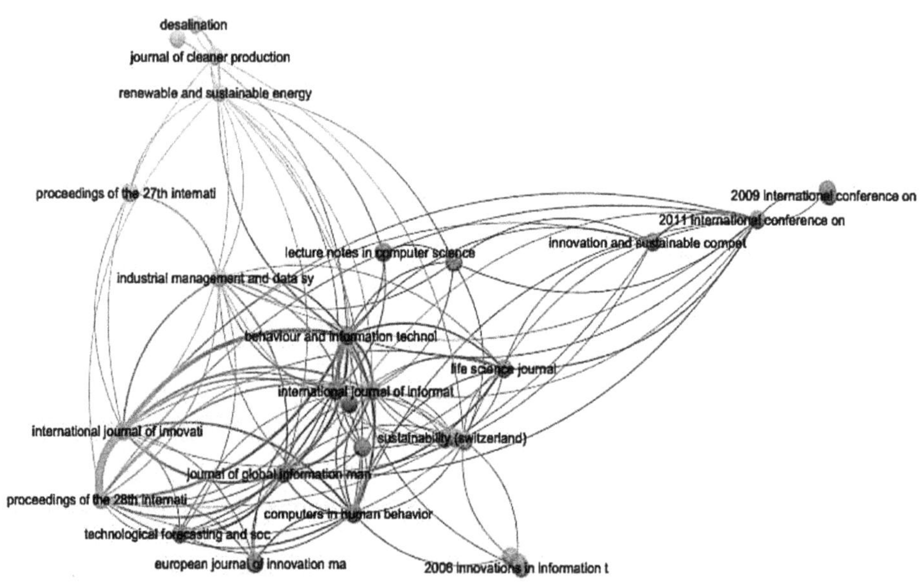

For a more accurate representation see the electronic version.

strongest link, indicating that they are mostly citing each other. International journal of innovation had also a strong link with behavior and information technology, computers in human behavior and international journal of information. It depicts that the researchers have been exploring various applications of computers and information technology.

Keyword Analysis of Publication From January 1990 to September 2018

This section of the chapter discusses keywords co-occurrence analysis of the research publications of GCC countries from January 1990 to September 2018. Keywords analysis is especially suitable to get the overview of a field as they show the main theme of the publication (Waaijer and Palmblad 2015). We analyzed the keyword co-occurrence in the articles published by the GCC countries from January 1990 to September 2018. The keyword co-occurrence in VOS viewer was set to a minimum of 3 occurrences.

It can be seen in figure 6 that there is a large number of keyword co-occurrences with knowledge economy which are; innovation, technology, development, education and knowledge management. At the same time, there is a recent outburst of the new key word for example; big data with the co-occurrences of terms; innovation, technology, knowledge management,

Figure 6. Keyword analysis of publication from January 1990 to September 2018

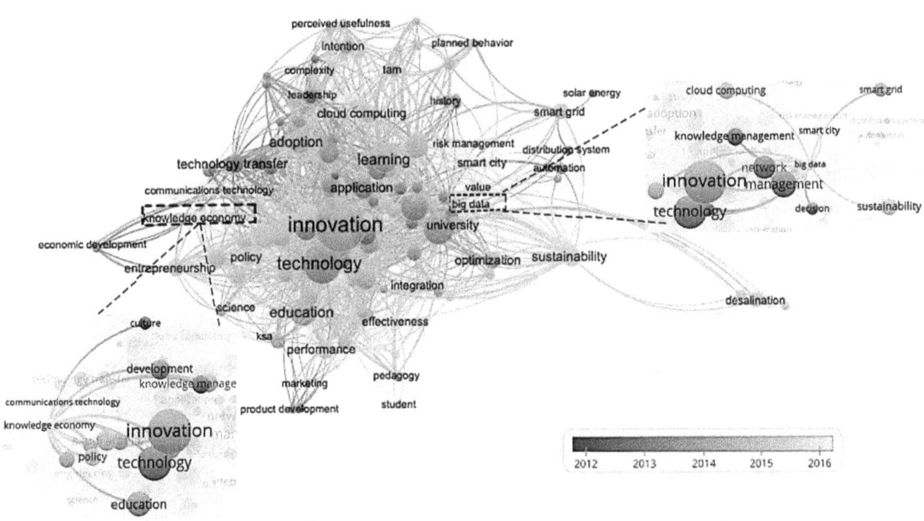

*For a more accurate representation see the electronic version.

cloud computing, smart city, and smart grid. The emergence of terms such as "smart city" and "smart grid" in the publications of GCC countries suggests that research opportunities are present in all of these domains.

CONCLUSION

This study shows that over the time period of 2008-2018 there is a huge increment in investing the resources in the GCC world to make move from just oil producing countries to knowledge-based economies. Optimistically, there is noticeable effect on GCC countries in publications output, citations count and international collaborations over this time period. The results of this research have following four outcomes.

First, Saudi Arabia has a high number of publication count and citation which shows that among all GCC countries Saudi Arabia is more concerned with the quality and influence of its research count. Moreover, the publication and citation count of GCC nations conforms to research expenditure on Research and Development (R&D) by these countries.

Second, publication productivity of GCC nations have been increased after 2004 and the most significant term in current years is innovation, which concluded that GCC countries are participating in research and innovation. We also note that the E-learning, knowledge management, cloud computing, education, technology etc. are the other frequent key terms in the publications of GCC countries and we found that these terms gained more attention of researcher in recent years. It implies that the researchers from GCC countries are also trying to expand the scope of innovation and technology in research and it is becoming important in the research field.

Third, most of the research by GCC countries is being done with the international collaborations and co-author publication of Saudi Arabia is high among all GCC countries. From these findings we can conclude that Saudi Arabia plays a dominating position of technological innovation aptitude in GCC, and USA, China, France, and United Kingdom are top countries of collaborative innovation with GCC. We found that the technological innovation competency of GCC countries has increased over time. However, GCC must focus on strengthening the intra- and int'l GCC collaborations due to the fact that they are effective at welcoming new ideas and sharing the costs of technological innovation.

Fourth, articles published by the GCC countries illustrate that opportunities are present in domains like: big data, innovation, technology and knowledge management. Last but not the least, the study helps to provide an insight to the research landscape of the GCC and useful information to the scientific community as well as to the technology and innovation policy makers for better decision making and to create network of excellence across GCC scientific power houses of research and development.

REFERENCES

Adams, J., King, C., Pendlebury, D., Hook, D., & Wilsdon, J. (2011). *Global research report Middle East exploring the changing landscape of Arabian, Persian and Turkish research.* Thomson Reuters.

Al-Jaber, H., & Dutta, S. (2008). *Qatar: Leveraging technology to create a knowledge-based economy in the Middle East.* Geneva, Switzerland: World Economic Forum.

Al-Yousif, Y. K. (2008). Education expenditure and economic growth: Some empirical evidence from the GCC countries. *Journal of Developing Areas, 42*(1), 69–80. doi:10.1353/jda.0.0025

Bahgat, G. (1999). Education in the Gulf monarchies: Retrospect and prospect. *International Review of Education, 45*(2), 127–136. doi:10.1023/A:1003610723356

Fan, P. (2011). Innovation capacity and economic development: China and India. *Economic Change and Restructuring, 44*(1-2), 49–73. doi:10.100710644-010-9088-2

Hassan, S. U., Sarwar, R., & Muazzam, A. (2016). Tapping into intra-and international collaborations of the Organization of Islamic Cooperation states across science and technology disciplines. *Science & Public Policy, 43*(5), 690–701. doi:10.1093cipolcv072

Karkouti, I. M. (2016). Qatar's educational system in the technology-driven era: Long story short. *International Journal of Higher Education, 5*(3), 182–189. doi:10.5430/ijhe.v5n3p182

Motohashi, K. (2005). University-industry collaborations in Japan: The role of new technology-based firms in transforming the national innovation system. *Research Policy, 34*(5), 583–594. doi:10.1016/j.respol.2005.03.001

Nasser, R., Cherif, M., & Romanowski, M. (2011). Factors that impact student usage of the learning management system in Qatari schools. *International Review of Research in Open and Distance Learning*, *12*(6), 39–62. doi:10.19173/irrodl.v12i6.985

Nasser, R., & Romanowski, M. (2011). Teacher perception of professional development in the context of national educational reform: The case of Qatar. *International Journal of Training and Development*, *15*(2), 158–168. doi:10.1111/j.1468-2419.2011.00377.x

Nelson, R. R. (1993). *National innovation systems: A comparative analysis*. New York: Oxford University Press.

Sabah, F., Hassan, S., Muazzam, A., Iqbal, S., Soroya, S. H., & Sarwar, R. (2018). Scientific collaboration networks in Pakistan and their impact on institutional research performance: A case study based on Scopus publications. *Library Hi Tech*. doi:10.1108/LHT-03-2018-0036

Sarwar, R., & Hassan, S. U. (2015). A bibliometric assessment of scientific productivity and international collaboration of the Islamic World in science and technology (S&T) areas. *Scientometrics*, *105*(2), 1059–1077. doi:10.100711192-015-1718-z

Shin, C., & Kim, Y. (2108). Changing patterns of higher education governance under Neoliberalism: Global and East Asian perspectives. In J. C. Shin (Ed.), *Higher education governance in East Asia* (pp. 223–241). Singapore: Springer Nature.

Stasz, C., Eide, E. E., Martorell, F., Constant, L., Goldman, C. A., Moini, J. S., ... Salem, H. (2007). *Post-secondary education in Qatar: Employer demand, student choice, and options for policy*. Santa Monica, CA: RAND Corporation.

Wiseman, A. W., & Anderson, E. (2012). ICT-integrated education and national innovation systems in the Gulf Cooperation Council (GCC) countries. *Computers & Education*, *59*(2), 607–618. doi:10.1016/j.compedu.2012.02.006

Wong, C. (2011). Rent-seeking, industrial policies and national innovation systems in Southeast Asian economies. *Technology in Society*, *33*(3-4), 231–243. doi:10.1016/j.techsoc.2011.09.003

Chapter 4
A Bibliometric Analysis of the Technological Innovation in China

Chien-wen Shen
National Central University, Taiwan

Jung-tsung Ho
National Central University, Taiwan

ABSTRACT

This chapter uses direct citation network analysis and bibliometric analysis approach to explore the overall research development of China's technological innovation. Results show that environmental science ecology, engineering, science technology, and energy fuels are the top research topics related to China. A direct citation network of technological innovation in China is also illustrated to show the major themes of the technological innovation among key researchers in China. In addition, the authors adopt the relatedness measure as the basis of cluster analysis, which classifies five main research development groups related to the technological innovation in China. They then review the core papers under each group to illustrate the detailed development of each group. Researchers and practitioners can refer the results to identify the key papers in each development group and understand the influential research that may lead to the important technological innovations in China.

DOI: 10.4018/978-1-5225-9012-5.ch004

INTRODUCTION

For the past 20 years, China becoming the "world factory" has changed the structure of the global trade market. During this process, in addition to dedicating on building its own production capability, China has not neglected to enhance its technological innovation capabilities (Altenburg, Schmitz, & Stamm, 2008), which has led to breakthrough development in many technological research fields. For example, the aerospace industry and the wind turbine industry, which are the development priorities under China's national support policies, have seen increased global market share and heightened research funding invested, and this has led to the wind tunnel testing technological researches flourished and closely follows after United States in recent years (Mo, Fu, & Ho, 2018). As China becomes the great political, economic and technology power, its reliance on foreign technology imports has also declined. Take low-carbon clean energy technology as an example, China's technological innovation capability has led to a technology transfer model, i.e. reversing itself from a major technology import country to an export country (Urban, 2018). Under this trend, China's technological innovation has gradually attracted the attention of the academic field, and its processes and problems were being explored from different scientific research areas. For example, eco-innovation drivers and their impact on business performance (Cai & Li, 2018), innovative technology developments in China's aircraft and space manufacturing (Yang, Hong, Song, & Hong, 2017) or companies use public venture capital funds to improve innovation and financial performance (Y.B. Wang, Li & Furman, 2017). Thousands of scientific research related papers were published reveals the importance of China's technological innovation, as well as providing the necessary data sets to explore the overall trends and details in the area of technological research and development.

Upon reviewing relevant research literature, it's evident that most of the papers on China's technological innovations start from specific industry or issue, such as logistics, renewable energy, and agricultural technology (Cui, Su, & Hertz, 2012; Huang et al., 2012; Z. Li & Zhang, 2013). As an emerging research area, there's hardly any paper taking a retrospective view on the overall development. Therefore, this study retrieved the relevant papers from the Social Citation Index Expanded and the Social Science Citation Index of the Web of Science database. From 1990 to 2018, a total of 1,875 articles were collected as the data set for our analysis. Then we explore the overall development of China's technological innovation researches using document analysis approach. The final results would help scholars to quickly establish a comprehensive and in-depth understanding in this area, and inspire them derive their own interest research topics. Since this study uses all the literatures in the database, it would be a highly labor-intensive and time-consuming project should we adopt an eye-reviewing approach to summarize the development of the overall scientific researches. Based on the fact that the authors of a paper should know the best which papers are most relevant to their research topics, we consider the direct citation as a critical analytical basis to extract the overall development direction of their research area. While direct citation in a paper represents the relative importance of papers in this field, commonly used measures like total citation number quantifies an individual scholar's overall citation impact. Here we use the direct citation in papers as the relationship network to show the major themes of the technological innovation among key researchers in China. We also adopt the relatedness measure as the basis of cluster analysis, which classifies five main research development groups related to the technological innovation in China. We then review the core papers under each group to illustrate the detailed development of each group. As we use algorithms to determine the five major development groups, and identify 60 key papers under each, researchers can directly find papers that are most relevant to their research topics from the key papers list. They can also refer our results to identify the key papers in each development group and understand the influential research that may lead to the important technological innovations in China.

TREND OF TECHNOLOGICAL INNOVATION IN CHINA

We retrieved and downloaded 1,875 research papers regarding Chinese technological innovation from the Web of Science database. Figure 1 shows the number of papers published by year. We found that before 2005 there were fewer papers related in this area. After 2005, the number of papers related to technological innovation gradually increased and most likely due to the gradual enhancement of China's technological innovation capabilities.

Figure 2 shows the quantitative distribution of research papers related to Chinese technological innovation in different scientific research areas. Among them, there are most papers published in the Business Economics area, with a total of 985 papers, accounting for 51.1% of all papers. It shows that the application of science and technological innovation has greatly

Figure 1. Number of papers by year

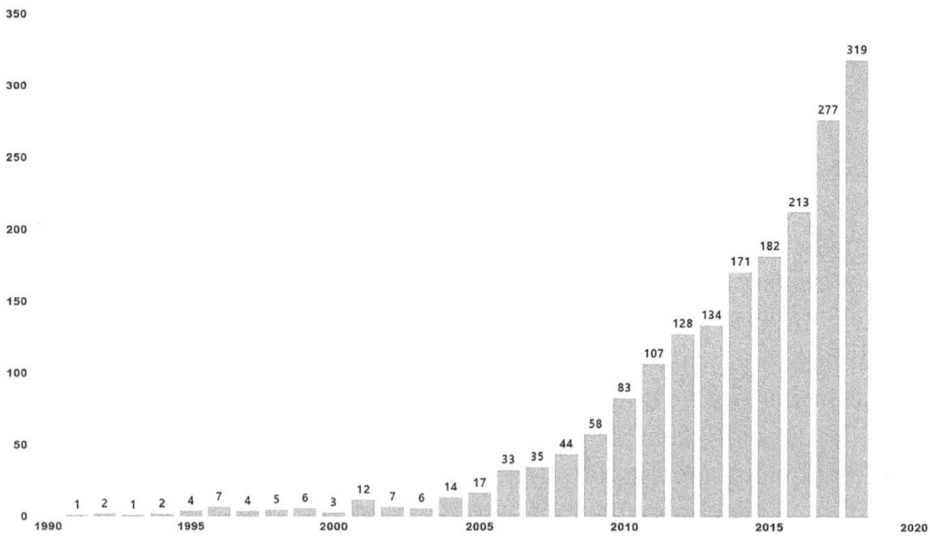

Figure 2. Number of papers by research area

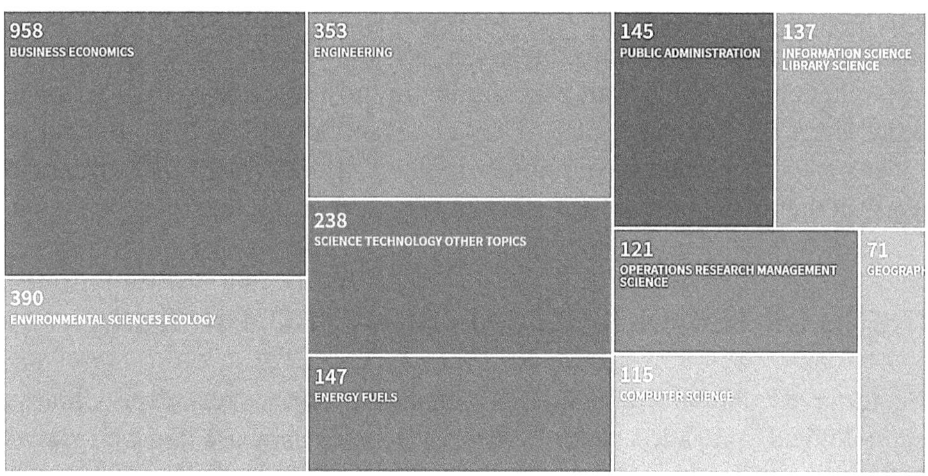

impacted China's business and economic environment. The top research areas are: Environmental Science Ecology (20.8%), Engineering (18.8%), Science Technology other Topic (12.7%) and Energy Fuels (7.8%). We can see that, by ranking the top ten different research fields, researches related to China's technological innovation spans across social science and engineering technology applications.

By looking at the research units, Figure 3 presents the top ten organizations in terms of the number of papers published in China's technological innovation. Among them, Tsinghua University published the biggest number of papers, a total of 105, accounting for 5.6% of the total papers. The following organizations are Chinese Academy of Sciences (4.8%), Zhejaing University (3.8%), University of the Chinese Academy of Sciences (2.2%) and City University of Hong Kong (2.1%). The top ten organizations are well-known universities and research institutions in China, and they have an edge in geographical advantage where data collection is handy. Besides, they are located at where the technological innovations are happening, any latest development grabs researchers' immediate attention.

RESULTS OF NETWORK ANALYSIS

Figure 4 shows the direct citation network relationship amongst Chinese technological innovation papers. The size of the node is in proportion with the number of edges of the node. The more edges, the larger the area of the nodes. From Figure 4 we can see that the largest node is labeled "li (2001a)" because it has the direct citation relationship with another 60 papers. This paper mainly focuses on the impact of product innovation strategies on the performance of Chinese technology start-ups (H. Y. Li & Atuahene-Gima, 2001). In addition, the second largest node labeled "altenburg (2008)" has

Figure 3. Number of papers by organization

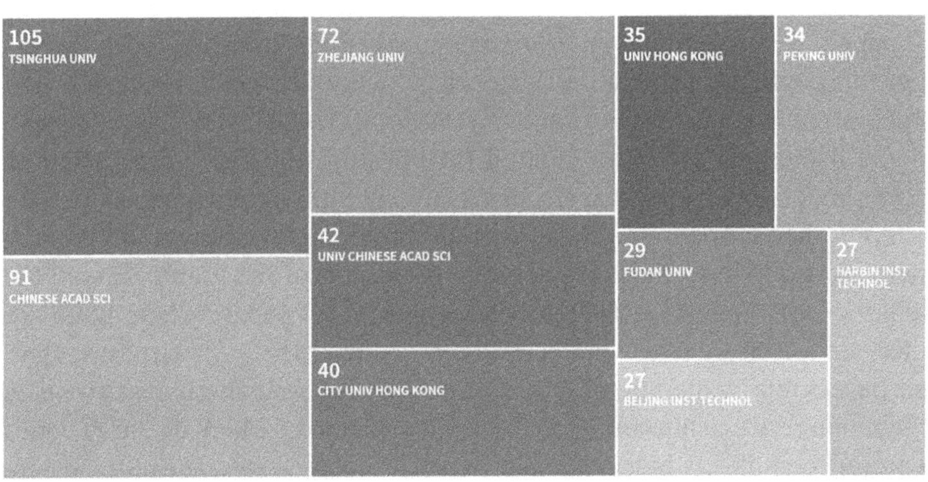

Figure 4. Direct citation network of technological innovation in China

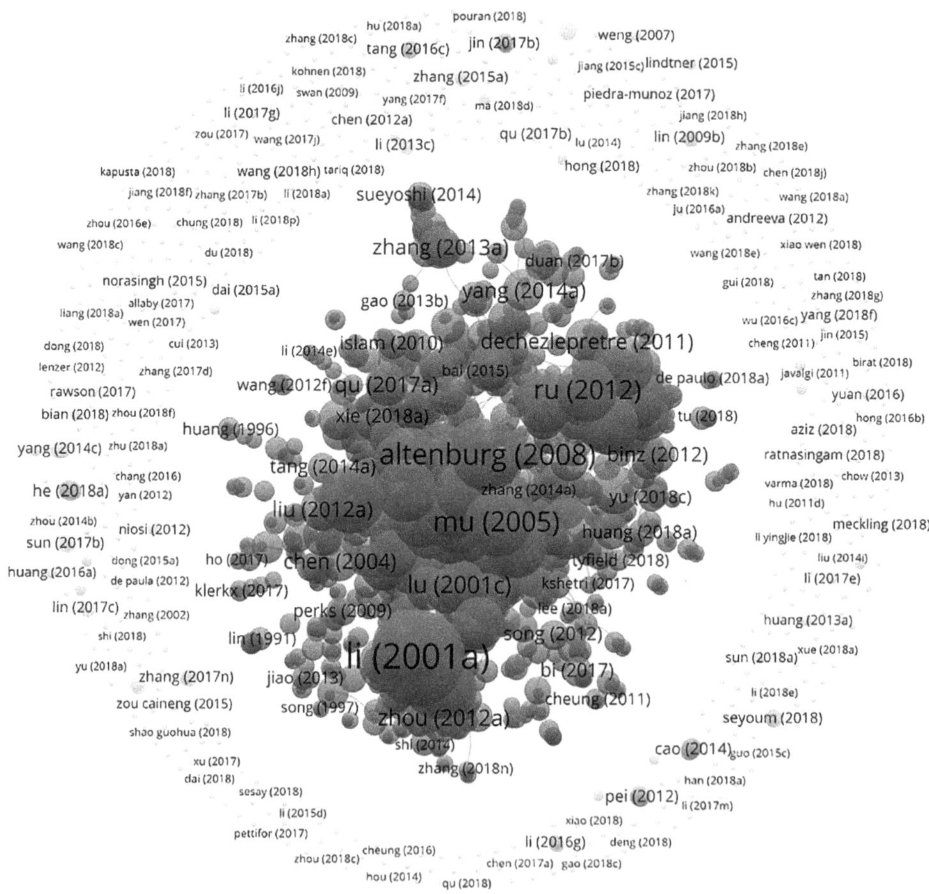

32 papers. The research topic is to explore whether China and India, having gone through the process of becoming "world factory" manufacturing process, can build up its own innovative capabilities (Altenburg et al., 2008). Figure 4 presents the basic citation relationship in China's scientific and technological innovation papers and relatively important papers. If we take a step further and examine closely the distribution of the papers in Figure 4, there are 1,068 blue nodes clustered in the center of the graph. These blue nodes are the largest citation network amongst all papers where there are inter-connecting citation with each other. At the graph's outskirt area, there are papers represented by gray and other color nodes. There are two kinds of citation relationships between these papers, among which 673 gray nodes represent completely independent papers without any reference relationship,

and 134 other color nodes represent the network groups where papers have 2-6 pieces of references. From the statistical analysis of all the papers, a total of 1,875 papers and total linked are 2,419, the average number of links per paper is 1.28, and the standard deviation is 3.48. From the nodes distribution and the statistics of the paper clusters in Figure 4, although the application of China's scientific and technological innovation is currently a popular research field, the current development is still relatively scattered, and a complete and systematic knowledge system is gradually being established.

In bibliometric research, we can use keywords that define specific topics to search for relevant papers. However, even if the content of the paper contains searched keywords, it may due to the multiple meanings of the words themselves, so we can't completely be certain that all the papers searched are related to the specific topics. In order to resolve this problem, we assume that if a single paper does not have a citation relationship with any other paper, we will conclude that the topics studied in the paper are not related to the topics studied in the research, so this paper should be excluded from subsequent analysis (Mejia & Kajikawa, 2017). Among the 1,875 papers in this research, there are 1,169 papers with more than one citation relationship, and 673 papers without mutual citation were excluded from the subsequent analysis. We identified the top 60 papers in China's technological innovation research field that have the most internal citation links: the papers with more links can be regarded as important papers in this research field. The cluster analysis shows that there are a total of 900 papers in the top five groups, accounting for 76.9% of the papers that have citation relationships. In order to focus on the mainstream of China's technological innovation research and development, our subsequent analysis uses the results of Group 1 to Group 5, while other smaller groups are not considered here.

Table 1 shows the top papers in Group 1, which include 371 papers. Among these papers, Y.F. Sun has a series of discussions on the development of technological innovation in China. Using patent to represent the level of innovation, he points out that China's patent rights concentrated in two major regions, namely, the coastal provinces with rapid economic development and the inland provinces with concentrated population. The paper argues that foreign companies' contribution to China's technological innovation is less than that of the local state-owned and collective-owned businesses (Sun, 2000). Another paper by Sun discusses the innovation behavior in terms of big and mid-size manufacturing enterprises. Similar conclusions show that the industrial innovation results developed by these enterprises' internal R&D teams are much greater than those developed from technologies

Table 1. Top cited papers in group 1

Author	Topic	Year	Internal Links
Xue, L	A historical perspective of China's innovation system reform: A case study	1997	19
Sun, YF	Spatial distribution of patents in China	2000	15
Lu, QW;	The organization of innovation in a transitional economy: Business and government in Chinese electronic publishing	2001	17
Sun, YF	Sources of innovation in China's manufacturing sector: Imported or developed in-house?	2002	24
Sun, Yf	China's national innovation system in transition	2002	12
Chang, PL	The innovation systems of Taiwan and China: A comparative analysis	2004	20
Mu, Q	Knowledge diffusion, market segmentation and technological catch-up: The case of the telecommunication industry in China	2005	26
Guan, JC	Collaboration between industry and research institutes/universities on industrial innovation in Beijing, China	2005	13
Zhou, Y	The making of an innovative region from a centrally planned economy: Institutional evolution in Zhongguancun science park in Beijing	2005	12
Eun, JH	Explaining the "university-run enterprises" in China: a theoretical framework for university-industry relationship in developing countries and its application to China	2006	23
Fan, Pl	Catching up through developing innovation capability: Evidence from China's telecom-equipment industry	2006	22
Liefner, I	Cooperation in the innovation process in developing countries: Empirical evidence from Zhongguancun, Beijing	2006	15
Tan, J	Growth of industry clusters and innovation: Lessons from Beijing Zhongguancun science park	2006	12
Chen, K	Universities/research institutes and regional innovation systems: The cases of Beijing and Shenzhen	2007	19
Altenburg, T	Breakthrough? China's and India's transition from production to innovation	2008	32
Hong, W	Decline of the center: The decentralizing process of knowledge transfer of Chinese universities from 1985 to 2004	2008	15
Kroll, H	Spin-off enterprises as a means of technology commercialisation in a transforming economy - Evidence from three universities in China	2008	14
Wang, CC	The growth and spatial distribution of China's ICT industry: New geography of clustering and innovation	2008	12
Guan, JC	Innovation strategy and performance during economic transition: Evidences in Beijing, China	2009	20
Shapira, P	From lab to market? Strategies and issues in the commercialization of nanotechnology in China	2009	10
Filatotchev, L	The export orientation and export performance of high-technology SMES in emerging markets: The effects of knowledge transfer by returnee entrepreneurs	2009	9
Sun, YF	Determinants of industrial innovation in China: evidence from its recent economic census	2010	12

continued on following page

Table 1. Continued

Author	Topic	Year	Internal Links
Hout, TM	China vs the world: Whose technology is it?	2010	11
Wang, CC	Industrial clustering and technological innovation in China: New evidence from the ICT industry in Shenzhen	2010	9
Liu, FC	China's innovation policies: Evolution, institutional structure, and trajectory	2011	26
Fu, XL	Indigenous and foreign innovation efforts and drivers of technological upgrading: Evidence from China	2011	24
Zhu, YM	Institution-based barriers to innovation in SMEs in China	2012	9
Wang, YD	The dual role of local sites in assisting firms with developing technological capabilities: Evidence from China	2013	9

directly obtained abroad. The main reason is that there's limited amount of effort paid to acquire these foreign technologies so creating serious hurdle for improving China's internal innovation capabilities (Sun, 2002b). Sun's another piece of retrospective paper reviewed through since 1950s shows that the nation's guidance is critical to China's innovation system development and performances (Sun, 2002a). Another case study paper also points out that state plays a critical role in the establishment of integrating governmental agencies' technological achievements with the innovation systems in corporate business activities (Lu & Lazonick, 2001). Another study by Sun reveals that foreign companies in China invested less R&D resources than domestic companies, while foreign companies' R&D activities were mostly concentrated in high-tech industries. The conclusion suggests that the Chinese government and domestic companies should focus on building their own innovation capabilities to attract more foreign companies to invest in R&D (Sun, 2010). In addition, the critical factors behind China Telecom's industrial technology capabilities as it took a leap from analog switch to electronic switch has also become a research topic. The results show that in addition to the right innovation strategy and knowledge diffusion from foreign companies to local ones, government's promotion and support to industry is also an important factor (Mu & Lee, 2005). The data results from 1,244 Chinese companies in Beijing show that corporate innovation activities were limited to domestic market and mainly focused on production quality improvement, furthermore Chinese manufacturing companies that received government support usually performed better (Guan, Yam, Tang, & Lau, 2009). Beijing Zhongguancun Science Park is considered to be the most innovative area in China. One research paper points out that during the process of technology upgrade and innovation,

ZGC benefited from technological collaboration with foreign companies as it helped generate new product ideas and made it into the market, while the collaboration with local universities mainly benefited new product design (Liefner, Hennemann, & Xin, 2006). Before the economic transformation in the mid-1990s, most of China's R&D resources were allocated to public research institutions/universities, so the impact of cooperation between industry and research institutions/universities has on technological innovation was also an important topic. A study of 950 enterprises analyzed the impact of this cooperation has on industrial innovation. The result shows that there's positive correlation between the technological novelty of industrial innovation and the cooperation with research institution/university. However, the same paper also points out that due to unstable cooperation relationship between enterprises and institutes, it's yet to achieve sufficient results to stimulate China's industrial innovation (Guan, Yam, & Mok, 2005). Another paper on the industrial characteristics of China's information and communication technology industry suggests that the ICT industry has a strong cluster effect in the coastal areas, but there is no significant relationship between this cluster effect and innovation performance, and the main reason is that distrust amongst companies has made it impossible to establish cross-company partnerships in innovation activities (CC Wang & Lin, 2008). The spin-off firms derived from universities are innovative models with Chinese characteristics. A study based on 82 interviews with spin-off firms shows that government-driven spin-offs have indeed become an appropriate solution for university technology transfer. However, many of the spin-off firms lacked performance, and the unique model of such university-derived companies has gradually been replaced by private companies that flourished after the reform and open-up (Kroll & Liefner, 2008). With strong local knowledge base seen in some developing countries and regions, one paper examines how local sites in developing countries can help their companies benefit from the spillover effects of international technology diffusion. The results of the study prove that China's local sites assist their companies in technological innovation in three indirect ways: local sites have enough local R&D personnel, cooperate with local universities and research institutions, and cooperate with local industrial community companies (YD Wang & Zhou, 2013).

Group 2 has a total of 179 articles and its top papers are summarized in Table 2. A study of the impact of product innovation strategies has on China's technology ventures performance found that the relationship between innovation and performance depends on environmental factors, including the environmental turbulence and institutional support, and the

Table 2. Top cited papers in group 2

Author	Topic	Year	Internal Links
Li, HY	Product innovation strategy and the performance of new technology ventures in China	2001	60
Li, HY	The adoption of agency business activity, product innovation, and performance in Chinese technology ventures	2002	20
Atuahene-Gima, K	Strategic decision comprehensiveness and new product development outcomes in new technology ventures	2004	9
Zhou, KZ	Innovation, imitation, and new product performance: The case of China	2006	18
Li, HY	The role of managers' political networking and functional experience in new venture performance: Evidence from China's transition economy	2007	32
Zhang, Y	Innovation search of new ventures in a technology cluster: The role of ties with service intermediaries	2010	31
Sheng, SB	The effects of business and political ties on firm performance: Evidence from China	2011	21
Zhou, KZ	How knowledge affects radical innovation: Knowledge base, market knowledge acquisition, and internal knowledge sharing	2012	14
Shu, CL	Managerial ties and firm innovation: Is knowledge creation a missing link?	2012	10

relationship-based strategies such as a strategic alliance established for product development. The research finding suggests that when exploring the performance of a product innovation strategy in a technology start-up, it is necessary to consider simultaneously the two strategic factors, the environmental-based and relationship-based ones, as moderators for analysis (HY Li & Atuahene-Gima, 2001). A start-up's relationship network also influences its technological innovation capability. From the perspective of industry cluster effect, a paper examines whether the partnership between start-ups with service intermediaries affects the ability to innovate. As service intermediaries have established extensive relationships with many companies, organizations and industries, the research suggests that start-ups should assimilate the network connection of service intermediaries, and leverage that to expand the scope of external innovation search and reduce search costs in order to enhance its innovation ability (Y. Zhang & Li, 2010). Another study clearly distinguishes the role of social tie in emerging economies as business relation and political relation. From the study results of 241 Chinese companies shows that business relationships have greater positive impact on performance than political relationships. When law enforcement is inefficient and technology changes rapidly, business relationship is more favorable (Sheng, Zhou, & Li, 2011). Another paper examines various types of management resources, such as political networks and functional experiences, and from a resource perspective, how to benefit start-ups in a transition economy. Survey data using new enterprises in Chinese high-tech industry as samples shows that managers' political relationship network and functional experience are positively correlated with new enterprises' performance (H. Y. Li & Zhang, 2007). The personal business and political relationships of upper-level managers can improve new enterprises' business performances. From the empirical data collected from 270 Chinese companies, it is found that the personal business relationship of upper-level managers has direct and huge impact on knowledge exchange and combination, while political relationship can only directly affect knowledge exchange. The functional role of political relationship is declining, while business relationship still imposes major impact on enterprise innovation for China in transition (Shu, Page, Gao, & Jiang, 2012). Another study compares the impact of innovation and imitation strategies on the performance of new products. Empirical evidence from cross-industry shows that innovation strategies can deliver better new product performance than imitation strategies. In addition, as market demand uncertainty increases, technology changes rapidly and competition intensifies, the benefits of innovation strategies over imitation strategies become more apparent (Zhou, 2006). Another study investigates the relationship between the breadth and depth of existing knowledge base and whether knowledge acquired through external market or internal knowledge sharing, to explore the impact of knowledge on technological innovation capabilities. The survey data from Chinese high-tech companies indicates that companies with wide extensive knowledge base are more likely to achieve radical innovation in the context of internal knowledge sharing. On the contrary, companies with deep knowledge base are better able to achieved radical innovation through knowledge acquired from external markets (Zhou & Li, 2012).

Group 3 has 141 papers and its top papers are summarized in Table 3. China has become the world's largest producer of solar photovoltaic cells in recent years, and the success drivers and limitations behind have attracted the attention from quite a few scholars. From the perspective of technology transfer and innovation, the key success factor of Chinese producers lies in the acquisition of the required technologies and skills through two major pathways: purchasing manufacturing equipment in highly competitive international markets and recruiting high-level R&D talents from foreign photovoltaic companies. But the results also point out that the success in the

Table 3. Top cited papers in group 3

Author	Topic	Year	Internal Links
Lewis, JI	Technology acquisition and innovation in the developing world: Wind turbine development in China and India	2007	24
De La Tour, A	Innovation and international technology transfer: The case of the Chinese photovoltaic industry	2011	23
Dechezlepretre, A	Invention and transfer of climate change-mitigation technologies: A global analysis	2011	13
Ru, P	Behind the development of technology: The transition of innovation modes in China's wind turbine manufacturing industry	2012	24
Grau, T	Survey of photovoltaic industry and policy in Germany and China	2012	10
Qiu, YM	The price of wind power in China during its expansion: Technology adoption, learning-by-doing, economies of scale, and manufacturing localization	2012	10
Gosens, J	From lagging to leading? Technological innovation systems in emerging economies and the case of Chinese wind power	2013	19
Lema, A	Technology transfer in the clean development mechanism: Insights from wind power	2013	12
Goodrich, AC	Assessing the drivers of regional trends in solar photovoltaic manufacturing	2013	9
Nahm, J	Scale-up nation: China's specialization in innovative manufacturing	2014	13
Zhang, F	Innovation and technology transfer through global value chains: Evidence from China's PV industry	2016	10

market is not reflected on the performance of technological innovation (de la Tour, Glachant, & Meniere, 2011). Another study compares industrial policy support for technology innovation in the photovoltaic industry in China and Germany. These policies include deployment supports, investment support for manufacturing plants and R&D support measures. Study result shows that although the development support is successful, both countries didn't link the support of manufacturing plants with incentives of innovation, while the support for R&D is relatively weak (Grau, Huo, & Neuhoff, 2012). Based on the historical photovoltaic price to evaluate the role of innovation technology in changing regional competitive advantage, one paper found that the historical price advantage of Chinese factories relative to US factories is not driven by country-specific advantages, but driven by scale and supply chain development. The result suggests that increasing the global installed capacity of photovoltaics from three aspects: technological innovation, manufacturing scale up and global cooperation (Goodrich, Powell, James, Woodhouse, & Buonassisi, 2013). Another paper explores how China has entered the global clean energy innovation system. Companies first entered photovoltaic module manufacturing through technology acquisitions, and then gradually built their global competitiveness through vertical integration strategies within the industry and a broader photovoltaic value chain. The main reason for the photovoltaic technology to transfer from global innovation system to China are: global market formation policies, talents international mobility, and flexibility of Chinese manufacturing and policy incentives by the Chinese government (F. Zhang & Gallagher, 2016). In addition, the market size of China's wind turbine manufacturing industry has grown substantially in recent years, and the technological innovation of this industry has also attracted the attention of scholars. China's innovation model began with imitating innovation, then turned to cooperative innovation, and more recently focused on achieving true independent innovation. From the retrospective of the transformation of innovation models and the dynamic interaction among technological capabilities, innovation models, market formation and wind energy policies, public policy is the key driver of the wind turbine industry innovation model evolvement and market development (Ru et al., 2012). Another study lists the system functions used to analyze the development of wind turbine technology through the Technology Information System framework, detailing the formation of China's wind turbine innovation system. The study found that China reduces its dependence on foreign knowledge and investment. It finally recommended that the industry development should focus on the quality of the wind turbine rather than on reducing costs. Meanwhile, it is necessary to establish a demonstration project for large-capacity and offshore wind turbine models to establish operational history and obtain quality certification (Gosens & Lu, 2013). International technology transfer is a key factor in ensuring that developing countries reduce carbon growth. Research indicates that the increasing technological capacity of host country has changed the nature of technology transfer, and the diversity of transfer mechanisms is influenced by pre-existing industrial and technological capabilities; researches and policies should pay more attention to the relationship between international low-carbon technology transfer mechanisms and local technological capabilities (Lema & Lema, 2013).

There are 104 papers are classified in Group 4, where the top papers are listed in Table 4. One paper from the knowledge flow perspective explores how the solar photovoltaic industry spread across from advanced technology countries to follower countries. This article used 19,105 solar photovoltaic patent database applications from Taiwan, South Korea and China retrieved from United States Patent and Trademark Office during 1984 to 2008, and

Table 4. Top cited papers in group 4

Author	Topic	Year	Internal Links
Wu, CY	Knowledge flows in the solar photovoltaic industry: Insights from patenting by Taiwan, Korea, and China	2012	12
Wang, ZH	Energy technology patents-co2 emissions nexus: An empirical analysis from China	2012	10

analyzed the knowledge flows in these patents using a set of 12 International Patent Classifications. The analysis found that these catch-up countries have increased their dependence on their own domestic knowledge generation and flow, showing that Taiwan, Korea and China have shifted from imitation to innovation strategies, and university research results are an important source of China's photovoltaic technology innovation (Wu & Mathews, 2012). Another study, from patent rights perspective, also explores the impact of energy technology innovation on reducing carbon emissions. By analyzing energy technology patents in 30 mainland China provinces from 1997 to 2008, the result shows that Sinopec fuel technology patents have no significant impact on CO2 emission reduction. However, domestic carbon-free energy technology patents seem to play an important role in reducing carbon dioxide emissions (ZH Wang, Yang, Zhang, & Yin, 2012).

Finally, Group 5 has 104 papers, where the top papers are summarized in Table 5. For China's technological start-ups, the R&D units established by

Table 5. Top cited papers in group 5

Author	Topic	Year	Internal Links
Mahmood, LP	Technological dynamism in Asia	2003	9
Von Zedtwitz, M	Managing foreign R&D laboratories in China	2004	28
Yam, RCM	An audit of technological innovation capabilities in Chinese firms: Some empirical findings in Beijing, China	2004	14
Chen, SH	Taiwanese IT firms' offshore R&D in China and the connection with the global innovation network	2004	11
Zhao, MY	Conducting R&D in countries with weak intellectual property rights protection	2006	14
Li, J	Product innovations in emerging economies: The role of foreign knowledge access channels and internal efforts in Chinese firms	2010	14
Zhang, Y	FDI spillovers in an emerging market: The role of foreign firms' country origin diversity and domestic firms' absorptive capacity	2010	11

foreign companies in China play a key role. One research paper specifically addresses the management of R&D units in foreign companies, by examining R&D tasks, site build-up, integration with superior levels and overall performance evaluation. The data came from an expert interview with 37 local R&D executives from 2001 to 2004, and it discusses the impact of culture on R&D management, local advantages, expatriate involvement and local laboratory organizations evolvement. The study found that the foreign R&D laboratories in China are not only an important carrier for local market development, but also an increasingly important source of local development technology (von Zedtwitz, 2004). Another study used data from 213 Chinese companies in Beijing China, by conducting regression analysis to examine the correlation between Technology Innovation Capability and innovation rate, sales growth and product competitiveness. The results confirm that R&D and resources allocation capabilities are the two most important TICs. Strong R&D capabilities can guarantee the innovation rate and product competitiveness for large and medium-sized enterprises, while resource allocation capability can increase the sales growth for small enterprises. In order to maintain sustainable development and improve overall innovation capabilities, Chinese companies should closely link TIC with technical strategy development and R&D innovation activities (Yam, Guan, Pun, & Tang, 2004). Another paper explores the impact of passive attitudes toward intellectual property protection on research and development capabilities in countries like China and India. The paper proposes that due to the weak intellectual property rights protection, it has resulted low return rate of innovation and low utilization of innovation talents. Therefore, multinational companies having ability to protect their intellectual property rights will find it attractive to conduct R&D at these locations. Multinational companies can establish internal R&D organizations to replace inappropriate external institutions, and take advantage of arbitrage between countries due to institutional gaps (Zhao, 2006).

CONCLUSION

This study uses direct citation network analysis and bibliometric analysis approach to explore the overall development of China's technological innovation research field, which can be classified into 5 groups. Group 1's main research topic is the interaction between technology innovation agents including state, local university research institutions, local companies and foreign companies. Among them, the state as a policy maker and a resource distributor, plays a pivotal role in the development of China's technological innovation capability. Group 2 explores the relationship between relationship network and the ability to build technological innovation. Scholars explores from different relationship network, including personal business relationship and political relationship of the company's managers and the industry cluster effect, on how to establish a knowledge exchange cooperative relationship. Most of the papers in Group 3 discuss the technological innovation in China's photovoltaic and wind turbine industry, trying to explain why China's grown by leaps and bounds in these two industries and the innovation challenges encountered. The papers in Group 4 discuss China's low-carbon clean energy technology. Several papers explore the flow and dissemination of its innovation capabilities by looking at the application of relevant patents at national or regional level. As there's direct relationship between technology innovation and R&D, so the papers under Group 5 investigate issues around R&D, such as management, unit establishment and the impact of cultural differences etc. Besides, R&D capabilities, product competitiveness and intellectual property rights protection are also important areas being reviewed. This study uses direct citation network approach to determine major themes of the technological innovation in China. By using the number of internal citations among the papers, we identified core papers under each development group. Researchers can directly refer to the core papers list and read the relevant papers in order to gain insight into the development of specific group, and explore research topics that are of particular interest to them, thus saving them a lot of time to filter out useful information from irrelevant papers.

REFERENCES

Altenburg, T., Schmitz, H., & Stamm, A. (2008). Breakthrough? China's and India's Transition from Production to Innovation. *World Development*, *36*(2), 325–344. doi:10.1016/j.worlddev.2007.06.011

Cai, W. G., & Li, G. P. (2018). The drivers of eco-innovation and its impact on performance: Evidence from China. *Journal of Cleaner Production*, *176*, 110–118. doi:10.1016/j.jclepro.2017.12.109

Cui, L. G., Su, S. L. I., & Hertz, S. (2012). Logistics Innovation in China. *Transportation Journal*, *51*(1), 98–117. doi:10.5325/transportationj.51.1.0098 de la Tour, A., Glachant, M., & Meniere, Y. (2011). Innovation and international technology transfer: The case of the Chinese photovoltaic industry. *Energy Policy*, *39*(2), 761–770. doi:10.1016/j.enpol.2010.10.050

Goodrich, A. C., Powell, D. M., James, T. L., Woodhouse, M., & Buonassisi, T. (2013). Assessing the drivers of regional trends in solar photovoltaic manufacturing. *Energy & Environmental Science*, *6*(10), 2811–2821. doi:10.1039/c3ee40701b

Gosens, J., & Lu, Y. L. (2013). From lagging to leading? Technological innovation systems in emerging economies and the case of Chinese wind power. *Energy Policy*, *60*, 234–250. doi:10.1016/j.enpol.2013.05.027

Guan, J. C., Yam, R. C. M., & Mok, C. K. (2005). Collaboration between industry and research institutes/universities on industrial innovation in Beijing, China. *Technology Analysis and Strategic Management*, *17*(3), 339–353. doi:10.1080/09537320500211466

Guan, J. C., Yam, R. C. M., Tang, E. P. Y., & Lau, A. K. W. (2009). Innovation strategy and performance during economic transition: Evidences in Beijing, China. *Research Policy*, *38*(5), 802–812. doi:10.1016/j.respol.2008.12.009

Huang, C., Su, J., Zhao, X. Y., Sui, J. G., Ru, P., Zhang, H. W., & Wang, X. (2012). Government funded renewable energy innovation in China. *Energy Policy*, *51*, 121–127. doi:10.1016/j.enpol.2011.08.069

Kroll, H., & Liefner, I. (2008). Spin-off enterprises as a means of technology commercialisation in a transforming economy - Evidence from three universities in China. *Technovation*, *28*(5), 298–313. doi:10.1016/j.technovation.2007.05.002

Lema, A., & Lema, R. (2013). Technology transfer in the clean development mechanism: Insights from wind power. *Global Environmental Change*, *23*(1), 301–313. doi:10.1016/j.gloenvcha.2012.10.010

Li, H. Y., & Atuahene-Gima, K. (2001). Product innovation strategy and the performance of new technology ventures in China. *Academy of Management Journal*, *44*(6), 1123–1134.

Li, H. Y., & Zhang, Y. (2007). The role of managers' political networking and functional experience in new venture performance: Evidence from China's transition economy. *Strategic Management Journal*, *28*(8), 791–804. doi:10.1002mj.605

Li, X., Wu, P., Shen, G. Q., Wang, X., & Teng, Y. (2017). Mapping the knowledge domains of Building Information Modeling (BIM): A bibliometric approach. *Automation in Construction*, *84*, 195–206. doi:10.1016/j.autcon.2017.09.011

Li, Z., & Zhang, H. P. (2013). Productivity Growth in China's Agriculture During 1985-2010. *Journal of Integrative Agriculture*, *12*(10), 1896–1904. doi:10.1016/S2095-3119(13)60598-5

Liefner, I., Hennemann, S., & Xin, L. (2006). Cooperation in the innovation process in developing countries: empirical evidence from Zhongguancun, Beijing. *Environment and Planning a-Economy and Space, 38*(1), 111-130.

Lu, Q. W., & Lazonick, W. (2001). The organization of innovation in a transitional economy: Business and government in Chinese electronic publishing. *Research Policy*, *30*(1), 55–77. doi:10.1016/S0048-7333(99)00082-7

Mejia, C., & Kajikawa, Y. (2017). Bibliometric analysis of social robotics research: Identifying research trends and knowledgebase. *Applied Sciences*, *7*(12), 1316. doi:10.3390/app7121316

Mo, Z., Fu, H.-Z., & Ho, Y.-S. (2018). Global development and trend of wind tunnel research from 1991 to 2014: A bibliometric analysis. *Environmental Science and Pollution Research International*, *25*(30), 30257–30270. doi:10.100711356-018-3019-6 PMID:30155635

Mu, Q., & Lee, K. (2005). Knowledge diffusion, market segmentation and technological catch-up: The case of the telecommunication industry in China. *Research Policy*, *34*(6), 759–783. doi:10.1016/j.respol.2005.02.007

Ru, P., Zhi, Q., Zhang, F., Zhong, X. T., Li, J. Q., & Su, J. (2012). Behind the development of technology: The transition of innovation modes in China's wind turbine manufacturing industry. *Energy Policy*, *43*, 58–69. doi:10.1016/j.enpol.2011.12.025

Sheng, S. B., Zhou, K. Z., & Li, J. J. (2011). The Effects of Business and Political Ties on Firm Performance: Evidence from China. *Journal of Marketing*, *75*(1), 1–15. doi:10.1509/jm.75.1.1

Shu, C. L., Page, A. L., Gao, S. X., & Jiang, X. (2012). Managerial Ties and Firm Innovation: Is Knowledge Creation a Missing Link? *Journal of Product Innovation Management*, *29*(1), 125–143. doi:10.1111/j.1540-5885.2011.00883.x

Sun, Y. F. (2000). Spatial distribution of patents in China. *Regional Studies*, *34*(5), 441–454. doi:10.1080/00343400050058693

Sun, Y. F. (2002a). China's national innovation system in transition. *Eurasian Geography and Economics*, *43*(6), 476–492. doi:10.2747/1538-7216.43.6.476

Sun, Y. F. (2002b). Sources of innovation in China's manufacturing sector: Imported or developed in-house? *Environment & Planning A*, *34*(6), 1059–1072. doi:10.1068/a34107

Sun, Y. F. (2010). Foreign research and development in China: A sectoral approach. *International Journal of Technology Management*, *51*(2-4), 342–363. doi:10.1504/IJTM.2010.033809

Urban, F. (2018). China's rise: Challenging the North-South technology transfer paradigm for climate change mitigation and low carbon energy. *Energy Policy*, *113*, 320–330. doi:10.1016/j.enpol.2017.11.007

van Eck, N. J., & Waltman, L. (2014). CitNetExplorer: A new software tool for analyzing and visualizing citation networks. *Journal of Informetrics*, *8*(4), 802–823. doi:10.1016/j.joi.2014.07.006

von Zedtwitz, M. (2004). Managing foreign R&D laboratories in China. *R & D Management*, *34*(4), 439–452. doi:10.1111/j.1467-9310.2004.00351.x

Wang, C. C., & Lin, G. C. S. (2008). The growth and spatial distribution of China's ICT industry: New geography of clustering and innovation. *Issues & Studies*, *44*(2), 145–192.

Wang, Y. B., Li, J. Z., & Furman, J. L. (2017). Firm performance and state innovation funding: Evidence from China's innofund program. *Research Policy*, *46*(6), 1142–1161. doi:10.1016/j.respol.2017.05.001

Wang, Y. D., & Zhou, Z. (2013). The dual role of local sites in assisting firms with developing technological capabilities: Evidence from China. *International Business Review*, *22*(1), 63–76. doi:10.1016/j.ibusrev.2012.02.003

Wang, Z. H., Yang, Z. M., Zhang, Y. X., & Yin, J. H. (2012). Energy technology patents-CO2 emissions nexus: An empirical analysis from China. *Energy Policy*, *42*, 248–260. doi:10.1016/j.enpol.2011.11.082

Wu, C. Y., & Mathews, J. A. (2012). Knowledge flows in the solar photovoltaic industry: Insights from patenting by Taiwan, Korea, and China. *Research Policy*, *41*(3), 524–540. doi:10.1016/j.respol.2011.10.007

Yam, R. C. M., Guan, J. C., Pun, K. F., & Tang, E. P. Y. (2004). An audit of technological innovation capabilities in Chinese firms: Some empirical findings in Beijing, China. *Research Policy*, *33*(8), 1123–1140. doi:10.1016/j.respol.2004.05.004

Yang, Y., Hong, J., Song, G., & Hong, S. (2017). Technology policy, technology strategy and innovation performance: Evidence from Chinese aircraft and spacecraft manufacturing. *Science & Public Policy*, *44*(5), 620–630.

Zhang, F., & Gallagher, K. S. (2016). Innovation and technology transfer through global value chains: Evidence from China's PV industry. *Energy Policy*, *94*, 191–203. doi:10.1016/j.enpol.2016.04.014

Zhang, Y., & Li, H. Y. (2010). Innovation search of new ventures in a technology cluster: The role of ties with service intermediaries. *Strategic Management Journal*, *31*(1), 88–109. doi:10.1002mj.806

Zhao, M. Y. (2006). Conducting R&D in countries with weak intellectual property rights protection. *Management Science*, *52*(8), 1185–1199. doi:10.1287/mnsc.1060.0516

Zhou, K. Z. (2006). Innovation, imitation, and new product performance: The case of China. *Industrial Marketing Management*, *35*(3), 394–402. doi:10.1016/j.indmarman.2005.10.006

Zhou, K. Z., & Li, C. B. (2012). How knowledge affects radical innovation: Knowledge base, market knowledge acquisition, and internal knowledge sharing. *Strategic Management Journal*, *33*(9), 1090–1102. doi:10.1002mj.1959

Section 2

Emerging Technologies and the Prospect of International Collaboration

Chapter 5
Blockchain Technologically Driven Innovation at the Gulf Cooperation Council Countries:
An Overview

Yousef Alabbasi
The University of New England, Australia

Kamaljeet Sandhu
The University of New England, Australia

ABSTRACT

This chapter delves into the innovation of blockchain technology and its benefits for advancing the economic growth in the Gulf Cooperation Council (GCC) countries. Blockchain technology provides the security and reliability that is required in the GCC region for enhancing the technology acceptance level of its population. Apart from providing security, blockchain technology offers financial efficiency, cost reduction, and the elimination of operational risks. Blockchain technology is equally beneficial for nonfinancial industries, such as the healthcare and education industry of the GCC region. The digital revolution based on blockchain technology is beneficial for GCC countries to shift their economic dependence from oil to innovations driven by technology. This will move the GCC to develop more smart cities as the GCC region has an ample amount of financial and human resources that are required for successful integration of blockchain technology in its numerous industries.

DOI: 10.4018/978-1-5225-9012-5.ch005

INTRODUCTION

Gulf Cooperation Council (GCC) is an economic hub in the Middle Eastern that comprises of six major Arab states, which include Saudi Arabia, Bahrain, Qatar, Oman, Kuwait and the United Arab Emirates. Among the Middle Eastern countries, only Iraq was excluded from the GCC establishment whereas inclusion of other countries, such as Morocco and Yemen, has been under consideration since few years now. The Charter of the Gulf Cooperation Council was signed by all member states on May 25 1981 (Ayuso, Villar, Pastor, & Fuentes, 2015). Since the establishment of the GCC charter, its member states have supported each other economically as well as politically for ensuring socioeconomic and socio-political sustainability in the region.

The GCC was mainly established for economic union, which facilitates the individual economic growth of the GCC states. It is further observed that the GCC countries have high dependence on their oil reserves, as Saudi Arabia, Kuwait and the United Arab Emirates are among the top countries with highest oil reserves on a global level. The total GDP of the GCC member states exceeds $3.5 trillion due to the large oil reserves and stable economic conditions in all GCC countries (Cammett, 2018). According to 2017 estimates, the total population of the GCC region has surpassed 54 million whereby foreign workers and expatriates represent a large percentage of the total population, especially in the United Arab Emirates. Accordingly, the economic growth of the GCC countries is majorly dependent on productivity of foreign workers, oil production, and foreign direct investment.

The increased dependence of GCC countries on their oil resources increases their risk exposure towards economic recession and financial downfalls. The drastic decrease in global oil prices during 2014 negatively affected the economic performance of GCC countries (Cammett, 2018; Al-Maamary, Kazem, & Chaichan, 2017). The governments of GCC countries have also previously realized the importance of shifting their dependence on economic efficiency from oil production. Over the last few years, GCC countries have explored numerous avenues for economic transformation. Among some of the greatest possibilities, digital technologies offer a promising economic future. The digital transformation suits the economic structure of GCC countries, especially because of abundance of capital and unsaturation of its different markets in terms of technological advancements and modifications. Despite the fact that the United Arab Emirates has taken a lead in digital transformation in the Middle Eastern region and among GCC member states, there are still major technological transformations that are required for establishing Dubai as the digital hub of the region.

Among different technological innovations and advancements in the 21st century, blockchain technology has gained popularity due to its countless benefits for the global IT industry as well as global financial industry. Blockchain technology induces automation in financial as well as nonfinancial transactions, allowing the client and institution to execute transactions without the interference of brokers (Paech, 2017). Such benefits of blockchain technology pose a high potential for maximizing the economic growth of GCC countries due to the elimination of financial risks from its banking industry, capital and stock markets, and information and communication technology (ICT) systems (Pinna, 2016). Irrespective of the large capital investment associated with induction of blockchain technology, it offers a thriving economic future for GCC countries.

BACKGROUND

Blockchain technology has been defined as, "blockchain is essentially a distributed database of records, or public ledger of all transactions or digital events that have been executed and shared among participating parties" (Crosby, Pattanayak, Verma, & Kalyanaraman, 2016, p. 8). Every transaction in the ledger is certified as far as majority of the participants in the system have the same recorded transaction. Accordingly, the reliability of a blockchain system is dependent on the linked transactions (blocks) whereby every transaction is connected to its preceding as well as proceeding transaction. In order to change a single block, the hackers or cybersecurity criminals need to change majority of the transactions (blocks) in the public ledger, which is theoretically impossible. *The Economist* has labeled blockchain technology as "the trust machine" because of its high reliability and minimum risk exposure (Crosby et al., 2016). The primary purpose of blockchain technology is in line with developing a trustful relationship between two parties while eliminating the role of brokers or intermediaries. This way, blockchain technology offers a relatively cheap and reliable method of performing transactions, storing data, and retrieving information. An economic system that is based on blockchain technology does not require human intervention, making it completely dependent on a verified algorithm that keeps a digital track of transactions (Iansiti & Lakhani, 2017). Any missing transaction or block in the overall public ledger is automatically rectified, as the blockchain system investigates every preceding and proceeding transaction.

There are financial as well as nonfinancial benefits of blockchain technology. For example, blockchain technology was first commercially used for offering a digital currency, Bitcoin, on a global level (Giungato, Rana, Tarabella, & Tricase, 2017). The huge success of blockchain technology resulted in global popularity of digital currencies, which led to the penetration of hundreds of digital currencies in the global financial industry. Digital currencies do not require the interference of brokers. Clients have the liberty to perform transactions on their own discretion without having the need to pay transaction fee to brokers. Eventually, banks as well as capital markets will be able to shift their systems to blockchain technology, which will maximize the efficiency of the global financial industry. On the other hand, blockchain technology is equally beneficial for nonfinancial industries. For example, blockchain technology can be used to set up electronic health records, which will offer higher data privacy and confidentiality to patients. Similarly, blockchain technology can be used for devising new applications for the education system. Apart from digital applications, blockchain technology is also capable of transforming traditional education platforms into digital education systems (Sharples & Domingue, 2016). In addition, blockchain technology also offers viable methods for handling monetary payments in the education industry, allowing students and educational institutes to connect with each other without depending on banks and other financial institutions.

BLOCKCHAIN TECHNOLOGY SIGNIFICANCE

The significance of blockchain technology is in line with its ability to offer greater reliability, cheaper transactions, and efficient systems. Also, blockchain technology has the capability to revolutionize global economy by enhancing the efficiency of numerous global industries (Tapscott & Tapscott, 2016). Blockchain technology is continuously expanding its roots in the global economy, gradually penetrating different global industries. Likewise, blockchain technology offers a promising economic future for GCC countries. The technology systems in GCC countries are comparatively less advanced than first world countries, such as in the United States and United Kingdom. According to Azar and Raouf (2017), GCC countries can use blockchain technology to improve the efficiency as well as reliability of their systems, which will help in increasing economic output, financial performance, and social sustainability. Accordingly, there are some elements affecting the blockchain acceptance and developing at GCC as following:

1. CFO Leadership Characteristics.
2. Culture.
3. Data Analytics.
4. Digital Training and Education.
5. System Quality and Security.

These elements will be discussed later in this chapter.

LITERATURE REVIEW

Blockchain as new technology at GCC, it will be subjected on two main theories: (Technology Acceptance Model) TAM and (Innovative Diffusion Theory) IDT.

TAM (Technology Acceptance Model)

The Technology Acceptance Model is a theory that evaluates the penetration power of new information systems or technologies. TAM is widely used to assess the acceptability of different information systems and technologies among consumers. According to Alharbi and Drew (2014), TAM has shown high validity on continuous basis, allowing it to become the "most widely-used framework" for measuring information system and technology acceptance of individuals. There are two major factors that fall under the domain of TAM and are used for analysing customer acceptability and decision-making. These factors that influence customer decision-making refer to "perceived usefulness" and "perceived ease-of-use" (Alharbi & Drew, 2014). Perceived usefulness measures the value addition of the information system or technology for the user, especially in increasing individual job performance. Accordingly, perceived usefulness directly as well as indirectly affects behavioural intentions of users. The indirect relationship between perceived usefulness and behavioural intentions is based on the intermediary variable of users' attitude towards using the information system or technology. On the other hand, perceived ease-of-use measures the effort required for the user to use the system (Marangunić & Granić, 2015).

Keeping in view the aforementioned factors of TAM, it is observed that blockchain technology has great potential for in-depth penetration among GCC residents. One of the major reasons for such findings refers to the high "perceived usefulness" of blockchain technology. Blockchain technology will not only increase the performance of individual users but will also enhance the quality of work performed. Similarly, utilization of TAM reveals that blockchain technology also has a higher perceived ease-of-use due to its focus on the algorithm and architecture of the information systems instead of influencing changes in the interface of information systems (Vo, Kundu, & Mohania, 2018). This means end users will keep on using the same application and information systems' interface while they will be able to shift from existing technologies to blockchain technology. Such ease of transformation to blockchain technology reflects high acceptance of blockchain technology in the GCC countries.

IDT (Innovative Diffusion Theory)

Innovative Diffusion Theory (IDT) or Diffusion of Innovations (DOI) aims at mesaruing the diffusion of technologies within a society. According to the IDT, an emerging or innovative idea takes some time to become acceptable among masses, and the IDT focuses on the motivations behind diffusion rate for a particular technology. The notion of masses has been used, as a new technology or innovative idea needs to be accepted by masses, in order to achieve socioeconomic sustainability (Azar & Raouf, 2017). The concept of IDT is based on five major categories, which are used to define the current stage of a new idea. These categories of IDT refer to innovators, early adopters, early majority, late majority, and laggards (Li, 2017). The market share of a new technology must reach laggards for the sake of maximizing its market share. The acceptance of early majority and late majority are the two most crucial stages for a new technology, as each stage represent 34% of the total market share.

From the perspective of IDT, blockchain technology is still in its first stage whereby the maximum potential of the target market is less than 2.5%. The market share of blockchain technology is much smaller in majority of the GCC countries due to lack of acceptance towards new technologies and inadequate public awareness (Alharbi & Drew, 2014). Among all the GCC countries, only Dubai has adopted a positive approach towards digital transformations based on blockchain technology. Still, Dubai and other states of the United Arab

Emirates have a long way to go when it comes to penetration of blockchain technology in its healthcare industry, education industry, and private sector (Sun, Yan, & Zhang, 2016). Dubai and some of the major cities of Saudi Arabia, Kuwait, and Bahrain are driving investment towards cryptocurrencies but the aforementioned countries does not own any cryptocurrencies (Sayed & Abbas, 2018). Accordingly, the IDT reflects blockchain technology at the innovators stage, which can move to early adopters and early majority in the next few years.

Blockchain as Digital Finance Platform (Cryptocurrencies)

The idea of digital finance system was introduced decades ago with the introduction of credit cards, wire transfers, and online banking accounts. With the passage of time, Internet has helped in revolutionizing the global financial system by establishing fintech services, such as Internet banking, mobile banking, and SMS banking (Chiu, 2017). Fintech helped in increasing the efficiency as well as penetration of financial services, allowing the global banking system to maximize its reach in terms of aiding international trade activities as well as helping the establishment of a digital finance system. The invention of blockchain technology helped in diffusion of cryptocurrencies on a global level. Cryptocurrencies are defined as, "a digital asset designed to work as a medium of exchange using cryptography to secure the transactions and to control the creation of additional units of the currency" (Osterrieder, Lorenz, & Strika, 2016, p. 58). Moreover, the utilization and application of cryptocurrencies has been explained as, "The underlying technology is based on a public ledger, or blockchain, shared between participants and a reward mechanism in terms of Bitcoins as an incentive for users to run the transaction network" (ElBahrawy, Alessandretti, Kandler, Pastor-Satorras, & Baronchelli, 2017, pp. 1-2). Bitcoin is the most popular cryptocurrency in the world with significant acceptance among the investors in GCC countries (Giungato et al., 2017). All types of cryptocurrencies fall under the domain of digital finance, which is a subdomain of fintech. Fintech investments, in terms of volume as well as number of deals, have been continuously increasing over the last 5 years, as shown in Figure 1.

Moreover, figure 2 shows the global fintech investment by region. It is evident that the global deal volume, including the rest of the world (ROW) region is showing consistent growth.

Figure 1. Fintech investments on a global level, 2012 – 2018

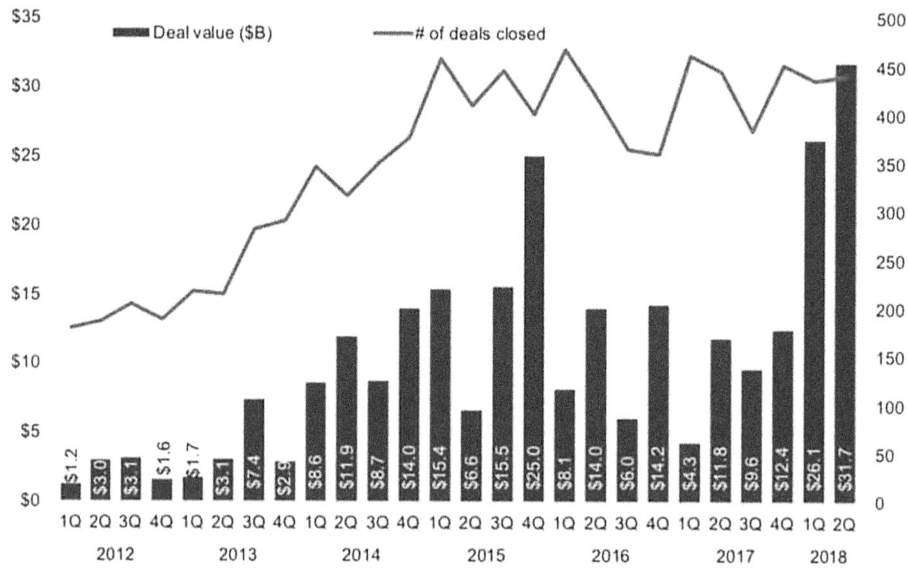

Figure 2. Global fintech financing activity by region

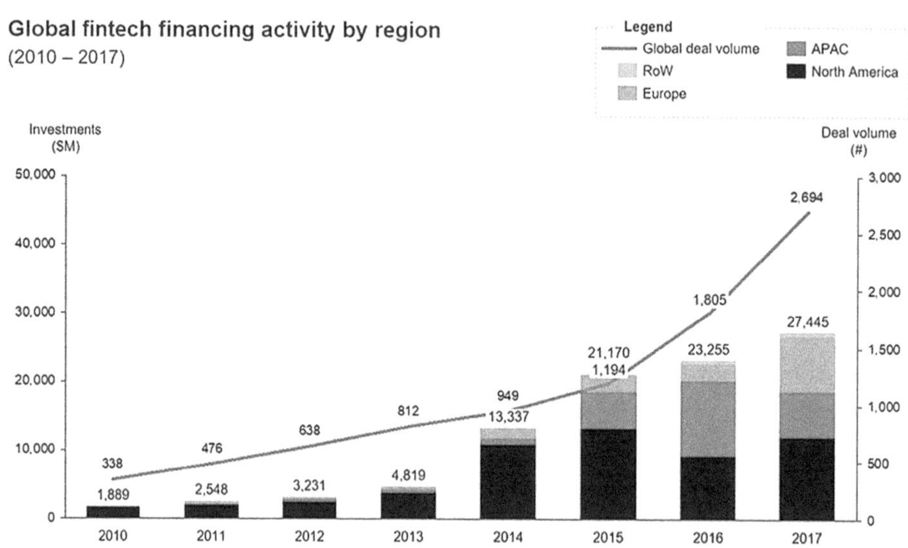

For a more accurate representation see the electronic version.

The abundance of financial capital in majority of the GCC countries has increased their acceptability towards induction of new technologies. Still, the healthcare system as well as the education system in GCC countries is relatively less advanced than first world countries like the United States and Australia. Accordingly, blockchain technology can increase the efficiency of health and education industries in GCC countries (Nguyen, 2016). Some of the steps may include use of blockchain technology for developing a reliable electronic health records database, applications that converge the gap between monetary payments and education system, and establishment of digital education and healthcare systems. Such initiatives require business innovation, in order to replace current technologies and traditional processes through blockchain technology and automated systems, respectively.

CFO Leadership and Characteristics

The Chief Financial Officer (CFO) is one of the most prominent positions in an organization. The CFO is directly responsible for making the financial decisions on behalf of the organization. Also, it is the ultimate responsibility of the CFO to manage assets, liabilities, equities, working capital, and cash flows of the organization, in order to consistently fund operations. It is observed that the main purpose of a CFO is to conduct financial planning, discuss it with senior management, and implement it with the help of financial officers and accountants. Keeping in view the commonalities between effective leadership traits and responsibilities of a CFO reveals that financial foresight, confidence, visionary skills, quick decision-making, and integrity are leadership characteristics of an efficient CFO (Fuessler, 2014; Atkinson, 2018). Such leadership characteristics help CFOs to make quick financial decisions while keeping in view the financial strengths and weaknesses of the organization. Moreover, confidence and visionary skills of CFOs help them grab financial opportunities that have a potential of yielding lucrative return on investments (Grant & Collings, 2016). Financial opportunities do not only include investments that generate direct return on investment but also include financial expenditure done for increasing the quality of systems, induction of information technology, and upgrading network architecture.

A good example of such financial investment refers to replacement of existing information systems and architecture with blockchain technology, in order to secure financial and economic future of the organization. It is the prime responsibility of CFO use personal perspective on risk before forwarding financial planning to the senior management (Dergel, 2014). Accordingly, if the induction of blockchain technology creates new avenues for achieving system efficiency and reliability while risks are eliminated, then the CFO should use effective communication skills to present the financial planning in front of the board. It is difficult for the senior management to approve an organizational change, new investment, or expense, unless it is financially viable (Fernández, Roscoe, & Aramaki, 2017). This means the implementation of blockchain technology should yield greater returns in the longer run when compared to its initial investment. This way, CFOs working for financial as well as nonfinancial companies in GCC countries has the responsibility to conduct thorough planning on the induction of blockchain technology within their own companies.

The feasibility report of using blockchain technology to replace existing technology systems has to project financial as well as economic benefits for the company. These benefits should focus on building a financially sustainable future for the company in the longer run. Blockchain technology may require additional capital investment initially for replacing existing technology systems, hiring new IT professionals, and training management employees for maximizing positive outcomes (Böhme, Chrisitin, Edelman, & Moore, 2015). The implementation of blockchain technology helps organizations to gain a competitive advantage, which has a tendency of strengthening brand image, creating loyal customers, and increasing revenues by multiple folds (Underwood, 2016; Khairuddin, Sas, Clinch, & Davies, 2016). The collection and presentation of aforementioned information falls under the job responsibilities of an effective CFO, who is willing to take an initiative for securing the financial future of the company.

Culture

Previous research on Arab cultural alignment with technology acceptance has yielded numerous limitations that resist technology penetration in Arab countries. On the other hand, the current population of the GCC countries represent a large percentage of youth population that comprises at least one-third to one-half of total GCC countries' population (AlMunajjed, Sabbagh, & Insight, 2011). According to a survey, at least 88% of this younger population likes to use Internet and interact with technology devices in their free time (AlMunajjed, Sabbagh, & Insight, 2011). Moreover, it is revealed that Arab countries have a strong stance on using their traditional communication styles and languages, which assures inadequate acceptance towards change (Al-Jumeily & Hussain, 2014; Omari, 2013). According to "Hofstede (1997) was amongst the first to investigate cultural dimensions and users' technology acceptance in Arab countries, and the findings revealed a high uncertainty-avoidance among Arab cultures, with people avoiding changes in their life—in this case, with regard to new computer technology" (Al-Jumeily & Hussain, 2014, p. 38). Still, the adoption rates of mobile phones in Arab countries reveal a contradictory story (Ameen & Willis, 2015).

Blockchain technology is one of the most advanced forms of technological advancements but it requires limited user awareness and training. Blockchain technology is limited to backend network and architecture of the information system instead of altering the influencing the technological interface of a computer software or mobile application (Meunier, 2018). Still, there is need of awareness in Arab culture, especially for middle classes, for understanding the concept and benefits of blockchain technology. Without such understanding, it would be difficult to motivate people for implementing or using blockchain technology-enabled services.

Data Analytics

Data analytics is among the most commonly used technologies in the world, at least since the advent of the 21st century. Some of the biggest companies like Google, Facebook, and Amazon use complex data analytics techniques to analyze consumer trends, consumer needs, popular products, flow of Internet traffic, and countless other variables (Gandomi & Haider, 2015). Data analytics is a process that allows examination and investigation of data sets for extracting information, which is then used to draw trends and conclusions. Data analytics require the use of specialized software and systems, as some data sets can be extremely large such as data sets involved in Big Data (Najafabadi et al., 2015). Companies like Wal-Mart and Amazon collect Big Data sets on daily basis that comprise of customers' financial information, common products, total purchasing amount, and even time of purchase.

On the other hand, some Big Data sets are so large in nature that it is not technically possible to analyze all the data contained in them. For example, if Facebook wants to analyze Big Data sets involving all the comments and likes that are made on Facebook in a single data, it is not possible to process such large amount of data on daily basis (Kambatla, Kollias, Kumar, & Grama, 2014). In comparison to data analytics systems and software being used on a global level, blockchain technology offers relatively easier data acquisition and sharing processes. This is also because of the elimination of middlemen and intermediaries, which makes it further difficult as well as slow to acquire, analyze, and share data. Numerous companies and IT specialists are working on the convergence of blockchain technology, big data analytics, and artificial intelligence, as it may help in revolutionizing the global data analytics industry (Tapscott & Tapscott, 2016). The convergence of such technologies will help in maximization of innovation and minimization of operational as well as compliance costs. The wide popularity of blockchain technology is also in line with the increased awareness of consumers regarding the need of transparency and trust while performing financial as well as non-financial transactions. According to Engelhardt (2017), healthcare organizations have started to use unique serial number coding using blockchain technology for tracking and locating medicines.

GCC countries are now increasingly relying on digital technologies for increasing their economic efficiencies. Dubai has emerged as one of the business hubs in the world, making it a great attraction for foreign diplomats and tourists. Gradually, Arabs are becoming more adaptive to a multicultural environment, which shows their positive approach towards acceptance and changes. Also, the United Arab Emirates along with Saudi Arabia and Kuwait are taking the lead in GCC countries for pursuing technology-based startups and innovative ideas (Sadayan & Rao, 2017). The aforementioned countries have influenced a digital drive in GCC region, allowing the government of every GCC country to increase their dependence on digital technologies. The digital drive has been initiated while viewing the benefits of data analytics and endless possibilities of artificial intelligence (Marinakis et al., 2018). The convergence of blockchain technology, big data analytics, and artificial intelligence might be a new concept for some of the GCC countries, but they are rapidly embracing future technologies. Such positive acceptance towards digital technologies will positively influence the integration of blockchain technologies with data analytics, as specialized systems and software are already being used in GCC countries in every domain.

Digital Training and Education

Training and education serve an important purpose in changing individual attitudes, behaviors and skills. Both of these knowledge enhancing tools, training and education, complement each other. Gradually, corporate organizations and multinationals have become dependent on training and development for continuously improving the skills of employees and managers. Also, corporate organizations and multinationals sponsor their employees and managers to acquire further education or participate in educational courses for enhancing their skills and expertise (Williams, 2018). One of the major reasons for such trends in the global corporate environment is the continuously changing technology that requires individuals to keep learning otherwise they are left behind.

Training and education are prerequisite for enhancing the skills and behavior of individuals. For this reason, training and education is closely associated with the implementation of blockchain technology (Mikroyannidis, Domingue, Bachler, & Quick, 2018). Without training and educating employees, it becomes extremely difficult for smooth implementation of new technology. Blockchain technology is still in its early phases of penetration in the global IT industry, despite its invention in 2007. The slow adaption of blockchain technology is in line with the lack of acceptance, its vast potential, and inadequate skills of majority of the IT professionals to integrate blockchain technology in modern systems (Alharbi & Drew, 2014). Researchers and IT professionals are slowly pursuing the possibilities of blockchain technology and are accordingly training and educating others for its possible potential. With the help of rigorous training and education, IT specialists can easily master their skills pertaining to blockchain technology. It is difficult to learn something when practical findings and results are not readily available. Blockchain technology is the main technology behind cryptocurrency industry and was strongly welcomed in the global healthcare industry. There are countless case studies, researches, and data analytics that can be easily used to learn familiarity with blockchain technology.

Similarly, training and education play a vital role in adoption of blockchain technology in GCC countries. Training and education require additional capital, which is in abundance in GC countries. The United Arab Emirates and Saudi Arabia are listed among the fastest growing emerging economies, with abundance of investment capital available with Kuwait and Bahrain governments (Callen, Cheris, Hasanov, Hegazy, & Khandelwal, 2014). This is the primary reason that GCC countries are adamant on joining their economic resources to induce digital innovations in their economies. Moreover, there is plenty of skilled labor in the Middle Eastern region due to amplified immigration activities in the region. Skillful individuals from Pakistan, India, China, Bangladesh, Sri Lanka, and other countries migrate to GCC countries for exploring economic opportunities (Kapiszewski, 2017). Such trends in the

Middle Eastern region have resulted in a high skilled employment industry in each of the GCC countries. Accordingly, it would be easy to train and educate IT professionals and managers for using blockchain technology. The two factors require for training and education refer to availability of capital and skilled workforce, both of which are readily available in GCC countries. Increased focus on training and education at the government level will ease the integration of digital technologies, which definitely include blockchain technology.

System Quality and Security

The security of blockchain technology represents its biggest advantage. The rapid popularity of Blockchain technology is strongly dependent on high level of security, reliability, and transparency it provides to its users. Blockchain has the ability to transform every public ledger that is in encrypted or non-encrypted form into a digital ledger with zero chances of error (Kshetri, 2017). The primary reason for such low operational risks refers to the architecture of blockchain technology. Transactions are documented digitally in blockchain technology but are connected to every proceeding as well as preceding transaction. These transactions are called blocks, and are connected to each other. There are millions and billions of blocks that may be connected to each other in a single blockchain system (Li & Xu, 2018). The reliability and security of blockchain technology has directly contributed towards its wide popularity. For instance, its use in digital currencies and wide array of other technology platforms has resulted in continuous revenue growth, as shown in the figure 3.

The blockchain algorithm automatically checks and verifies any false entries. The false entries are detected due to their lack of coherence with proceeding or preceding entries. The blockchain system removes and corrects the entries on the basis of a single rule, which is to believe what majority of the blocks, entries, or transactions represent (Matsuo, 2017). This means the blockchain system would remove and rectify even if up to 49% of the entries in the digital ledger have errors. The blockchain system cannot result in error until or unless more than 50% entries are incorrectly coded, which practically refer to millions of computers and users. For this reason, it is technically impossible that a blockchain system makes an error.

The cultural and social values of Arab nations are inclined towards traditional beliefs, as they have a high level of uncertainty avoidance index

Figure 3. Blockchain regional revenue generation, 2016 – 2025 (projected)

**For a more accurate representation see the electronic version.*

(Duran et al., 2016). The uncertainty avoidance index is one of the dimensions in the Hofstede's cultural model. This particular cultural dimension measures the ability of the people to be comfortable with ambiguity and uncertainty (De Meulenaer, De Pelsmacker, & Dens, 2015). A high level of uncertainty avoidance index among Arab people is an indication that Arabs do not welcome ambiguity and uncertainty. Such cultural and social values of Arabs have limited their exposure towards digital technologies. A major reason, for such cultural and social values, is based on the lack of transparency and trust issues that are involved in the current global financial system (Baptista & OIiveira, 2015). If the transparency and trust issues were removed from the equation, then it would become much easier for the Arabs to embrace digital technologies. The integration of blockchain technology can help the Arab world to overcome their cultural barriers, and enhance their technology acceptance level.

Offering greater reliability, increasing efficiency, reduced compliance costs, and minimized operations risks are some of the basic features of blockchain technology. The aforementioned features of blockchain technology have been developed while keeping in view the challenges faced by the global financial system. Despite cybersecurity measures, hackers are able to retrieve private and confidential financial information (Khan & Salah, 2018). Financial transactions are often intercepted, and international regulatory authorities have failed to overcome money-laundering activities. Blockchain technology offers a viable solution to all these challenges that prevail in the global financial industry (Till, Peters, Afshar, & Meara, 2017). Such effectiveness of blockchain technology offers a great potential in the GCC region whereby investors demand maximum financial security and transparency level.

FUTURE WORK

The researches of Blockchain technology is still too small and insufficient. That means more work on the Blockchain governance and adoption needs to take a place, Since the United Arab Emirates has taken the lead to initiate blockchain technology-based global trade activities. The United Arab Emirates along with Saudi Arabia can pursue a joint collaboration with China for developing a blockchain-based financial system that eliminates the risk involved in global trade. Such initiatives will not only increase the efficiency of global trade but will increase the concentration of new technology development in GCC region. As technological driven economy, the gap of lacking research must be taking in order to adopt the technology in both public sector and private sector. Moreover, blockchain technology offers a vast potential for revolutionizing the healthcare industry as well as education sector in GCC region (Tapscott & Tapscott, 2016). The aforementioned industries carry significance importance with respect to the creation of a welfare state. This means blockchain technology has the potential to bridge the socioeconomic gap between first world countries and emerging economics in GCC region. Furthermore, pursuing excellence in blockchain technology may drive further foreign direct investment in GCC region, which will help in growing its IT industry. New jobs, higher consumer spending, and increased tourism activities are among the positive outcomes of pursuing blockchain technology.

CONCLUSION

Blockchain technology is the future of global IT industry. The drive towards digital technologies in GCC region should be based on blockchain technology for gaining a competitive advantage on a global level. Apart from the financial industry, blockchain technology can help GCC countries to improve the efficiency of their healthcare industry, education industry, and private sector.

In addition, blockchain technology is beneficial for reducing operational risks and minimizing compliance costs. Such benefits of blockchain technology pose a high potential for maximizing the economic growth of GCC countries due to the elimination of financial risks from its banking industry, capital and stock markets, and information and communication technology (ICT) systems. Also, the strong dependence of GCC countries on oil reserves creates the need of digital evolution in the region.

An economic system that is based on blockchain technology does not require human intervention, making it completely dependent on a verified algorithm that keeps a digital track of transactions. Accordingly, the integration of blockchain technology can help the Arab world to overcome their cultural barriers and enhance their technology acceptance level. Based on TAM main factors, blockchain technology has a high "perceived usefulness" value as well as high "perceived easy-to-use" value. Furthermore, the convergence of blockchain technology, big data analytics, and artificial intelligence also offers a new avenue of technological innovation and growth, which may provide the digital transformation required in the GCC region for minimizing their dependence on oil production. In order to maximize the integration of blockchain technology, effective leadership and training of CFOs and IT professionals is crucial due to their significance in the implementation of new technology-based initiatives.

REFERENCES

Al-Jumeily, D., & Hussain, A. (2014). The impact of cultural factors on technology acceptance: A technology acceptance model across eastern and western cultures. *International Journal of Enhanced Research in Educational Development*, *4*(2), 37–62.

Al-Maamary, H. M., Kazem, H. A., & Chaichan, M. T. (2017). The impact of oil price fluctuations on common renewable energies in GCC countries. *Renewable & Sustainable Energy Reviews*, *75*, 989–1007. doi:10.1016/j.rser.2016.11.079

Alharbi, S., & Drew, S. (2014). Using the technology acceptance model in understanding academics' behavioural intention to use learning management systems. *International Journal of Advanced Computer Science and Applications*, *5*(1), 143–155. doi:10.14569/IJACSA.2014.050120

Alharbi, S., & Drew, S. (2014). Using the technology acceptance model in understanding academics' behavioural intention to use learning management systems. *International Journal of Advanced Computer Science and Applications*, *5*(1), 143–155. doi:10.14569/IJACSA.2014.050120

AlMunajjed, M., Sabbagh, K., & Insight, I. C. (2011). *Youth in GCC countries: meeting the challenge.* Booz & Company Inc.

Ameen, N., & Willis, R. (2015). The effect of cultural values on technology adoption in the Arab countries. *International Journal of Information Systems, 2.*

Atkinson, A. (2018). Leadership Is Not Enough. In *Finance Unleashed* (pp. 57–62). Cham: Palgrave Macmillan. doi:10.1007/978-3-319-66370-8_9

Ayuso, A., Villar, S., Pastor, C., & Fuentes, M. (2015). Actors and Opportunities; Interregional Processes in the Arab Region and Latin America and the Caribbean. *Atlantic Future Scientific Paper, 25.*

Azar, E., & Raouf, M. A. (2017). Sustainability issues in the GCC. In *Sustainability in the Gulf* (pp. 27–30). Routledge. doi:10.4324/9781315174884-1

Baptista, G., & Oliveira, T. (2015). Understanding mobile banking: The unified theory of acceptance and use of technology combined with cultural moderators. *Computers in Human Behavior*, *50*, 418–430. doi:10.1016/j.chb.2015.04.024

Böhme, R., Christin, N., Edelman, B., & Moore, T. (2015). Bitcoin: Economics, technology, and governance. *The Journal of Economic Perspectives*, *29*(2), 213–238. doi:10.1257/jep.29.2.213

Callen, M. T., Cherif, R., Hasanov, F., Hegazy, M. A., & Khandelwal, P. (2014). *Economic diversification in the GCC: Past, present, and future*. International Monetary Fund.

Cammett, M. (2018). *A political economy of the Middle East*. Routledge. doi:10.4324/9780429492600

Chiu, I. H. (2017). A new era in fintech payment innovations? A perspective from the institutions and regulation of payment systems. *Law. Innovation and Technology*, *9*(2), 190–234.

Crosby, M., Pattanayak, P., Verma, S., & Kalyanaraman, V. (2016). Blockchain technology: Beyond bitcoin. *Applied Innovation*, *2*, 6–10.

De Meulenaer, S., De Pelsmacker, P., & Dens, N. (2015). Have no fear: How individuals differing in uncertainty avoidance, anxiety, and chance belief process health risk messages. *Journal of Advertising*, *44*(2), 114–125. doi:10.1080/00913367.2015.1018465

Dergel, S. (2014). *Guide to CFO success: Leadership strategies for corporate financial professionals*. John Wiley & Sons. doi:10.1002/9781118691663

Duran, M., Irfan, S., Sipoche, D., Blough, D., Turner, D., Nguyen, P., & Steinberg, H. (2016). Customer Service & Hofstede's Cultural Dimensions Among Arabs, Australians, Canadians, Greek, Jamaicans, Pakistani, Singaporian, & American Accounting Information Personnel. *Allied Academies International Conference. Academy of Accounting and Financial Studies Proceedings*, *21*(1), 6.

ElBahrawy, A., Alessandretti, L., Kandler, A., Pastor-Satorras, R., & Baronchelli, A. (2017). Evolutionary dynamics of the cryptocurrency market. *Royal Society Open Science*, *4*(11), 170623. doi:10.1098/rsos.170623 PMID:29291057

Engelhardt, M. A. (2017). Hitching healthcare to the chain: An introduction to blockchain technology in the healthcare sector. *Technology Innovation Management Review*, *7*(10).

Fernández-Aráoz, C., Roscoe, A., & Aramaki, K. (2017). Turning Potential into Success: The Missing Link in Leadership Development. *Harvard Business Review*, 86–93.

Fuessler, B. (2014). "Performance Accelerators" push the frontiers of CFO leadership. *Strategy and Leadership*, *42*(3), 40–46. doi:10.1108/SL-03-2014-0029

Gandomi, A., & Haider, M. (2015). Beyond the hype: Big data concepts, methods, and analytics. *International Journal of Information Management*, *35*(2), 137–144. doi:10.1016/j.ijinfomgt.2014.10.007

Giungato, P., Rana, R., Tarabella, A., & Tricase, C. (2017). Current Trends in Sustainability of Bitcoins and Related Blockchain Technology. *Sustainability*, *9*(12), 2214. doi:10.3390u9122214

Grant, G. G., & Collins, R. (2016). The Role of Leadership. In *The Value Imperative* (pp. 173–191). New York: Palgrave Macmillan.

Iansiti, M., & Lakhani, K. R. (2017). The truth about blockchain. *Harvard Business Review*, *95*(1), 118–127.

Kambatla, K., Kollias, G., Kumar, V., & Grama, A. (2014). Trends in big data analytics. *Journal of Parallel and Distributed Computing*, *74*(7), 2561–2573. doi:10.1016/j.jpdc.2014.01.003

Kapiszewski, A. (2017). Arab versus Asian migrant workers in the GCC countries. In *South Asian Migration to Gulf Countries* (pp. 66–90). Routledge India.

Khairuddin, I. E., Sas, C., Clinch, S., & Davies, N. (2016, May). Exploring motivations for bitcoin technology usage. In *Proceedings of the 2016 CHI Conference Extended Abstracts on Human Factors in Computing Systems* (pp. 2872-2878). ACM. 10.1145/2851581.2892500

Khan, M. A., & Salah, K. (2018). IoT security: Review, blockchain solutions, and open challenges. *Future Generation Computer Systems*, *82*, 395–411. doi:10.1016/j.future.2017.11.022

Kshetri, N. (2017). Can blockchain strengthen the internet of things? *IT Professional*, *19*(4), 68–72. doi:10.1109/MITP.2017.3051335

Li, E. Y. (2017). *Integrating Innovation Diffusion Theory and the Technology Acceptance Model: The adoption of blockchain technology from business managers' perspective*. Academic Press.

Liu, L., & Xu, B. (2018, April). Research on information security technology based on blockchain. In *2018 IEEE 3rd International Conference on Cloud Computing and Big Data Analysis (ICCCBDA)* (pp. 380-384). IEEE. 10.1109/ICCCBDA.2018.8386546

Marangunić, N., & Granić, A. (2015). Technology acceptance model: A literature review from 1986 to 2013. *Universal Access in the Information Society*, *14*(1), 81–95. doi:10.100710209-014-0348-1

Marinakis, V., Doukas, H., Tsapelas, J., Mouzakitis, S., Sicilia, Á., Madrazo, L., & Sgouridis, S. (2018). From big data to smart energy services: An application for intelligent energy management. *Future Generation Computer Systems*. doi:10.1016/j.future.2018.04.062

Matsuo, S. I. (2017, October). How formal analysis and verification add security to blockchain-based systems. In *Proceedings of the 17th Conference on Formal Methods in Computer-Aided Design* (pp. 1-4). FMCAD Inc. 10.23919/FMCAD.2017.8102228

Meunier, S. (2018). Blockchain 101: What is Blockchain and How Does This Revolutionary Technology Work? In Transforming Climate Finance and Green Investment with Blockchains (pp. 23-34). Academic Press.

Mikroyannidis, A., Domingue, J., Bachler, M., & Quick, K. (2018, June). A Learner-Centred Approach for Lifelong Learning Powered by the Blockchain. In EdMedia+ Innovate Learning (pp. 1388-1393). Association for the Advancement of Computing in Education (AACE).

Najafabadi, M. M., Villanustre, F., Khoshgoftaar, T. M., Seliya, N., Wald, R., & Muharemagic, E. (2015). Deep learning applications and challenges in big data analytics. *Journal of Big Data*, *2*(1), 1. doi:10.118640537-014-0007-7

Nguyen, Q. K. (2016, November). Blockchain-a financial technology for future sustainable development. In *Green Technology and Sustainable Development (GTSD), International Conference on* (pp. 51-54). IEEE. 10.1109/GTSD.2016.22

Omari, A. (2013). Technology adoption in the Arabian Gulf countries: The case of e-Government. *IJCSEIT*, *3*(3), 1–8. doi:10.5121/ijcseit.2013.3301

Osterrieder, J., Lorenz, J., & Strika, M. (2016). Bitcoin and cryptocurrencies-not for the faint-hearted. *International Finance and Banking*, *4*(1), 56–94. doi:10.5296/ifb.v4i1.10451

Paech, P. (2017). The governance of blockchain financial networks. *The Modern Law Review*, *80*(6), 1073–1110. doi:10.1111/1468-2230.12303

Pinna, A. (2016). Distributed ledger technologies in financial markets?: An introduction and some points of interest for legal analysis. Chapter II—Legal developments in the ESCB central banking functions, 109, 117.

Sadayan, S. M. K., & Rao, K. (2017). Digital Inclusion Index: A Measurement of ICT Advancements in Bahrain & GCC. *International Journal of Computing and Network Technology*, *5*(02), 59–64. doi:10.12785/ijcnt/050203

Sayed, M. N., & Abbas, N. A. (2018). Impact of crypto-currency on emerging market focus on gulf countries. *Life Science Journal*, *15*(1).

Sharples, M., & Domingue, J. (2016, September). The blockchain and kudos: A distributed system for educational record, reputation and reward. In *European Conference on Technology Enhanced Learning* (pp. 490-496). Springer. 10.1007/978-3-319-45153-4_48

Sun, J., Yan, J., & Zhang, K. Z. (2016). Blockchain-based sharing services: What blockchain technology can contribute to smart cities. *Financial Innovation*, *2*(1), 26. doi:10.118640854-016-0040-y

Sweidan, O. D., & Alwaked, A. A. (2016). Economic development and the energy intensity of human well-being: Evidence from the GCC countries. *Renewable & Sustainable Energy Reviews*, *55*, 1363–1369. doi:10.1016/j.rser.2015.06.001

Tapscott, D., & Tapscott, A. (2016). *Blockchain revolution: how the technology behind bitcoin is changing money, business, and the world*. Penguin.

Till, B. M., Peters, A. W., Afshar, S., & Meara, J. G. (2017). From blockchain technology to global health equity: can cryptocurrencies finance universal health coverage? *BMJ Global Health*, *2*(4), e000570.

Underwood, S. (2016). Blockchain beyond bitcoin. *Communications of the ACM*, *59*(11), 15–17. doi:10.1145/2994581

Vo, H. T., Kundu, A., & Mohania, M. K. (2018). *Research Directions in Blockchain Data Management and Analytics*. EDBT.

Williams, P. (2018). Does competency-based education with blockchain signal a new mission for universities? *Journal of Higher Education Policy and Management*, 1–14.

Chapter 6
Technologically Driven Legal Framework of Blockchain and Cryptocurrencies

Ahmed Ashoor
University of New England, Australia

Kamaljeet Sandhu
University of New England, Australia

ABSTRACT

Blockchain technology refers to a digital, immutable, distributed ledger that registers completed transactions in a well-ordered manner and near real time. Blockchain security creates a decentralized environment that bars any third-party organization from controlling the cryptographically validated transactions and data. Blockchain technology fosters business innovation by creating a peer-to-peer networking that prevents one central server from accessing as well as processing data belonging to all companies in the network. Cryptocurrency can be defined as a digital asset built to facilitate completed transactions using cryptography. It helps in providing protection to the completed transactions and controlling the creation of additional units of the currency. In the recent years, the application of blockchain technology has been associated with governance. Blockchain governance has been applied in different fields; for example, it can be used to create permanent laws that cannot be violated by any third party.

DOI: 10.4018/978-1-5225-9012-5.ch006

INTRODUCTION

In the contemporary society, the internet has created a virtual community where billions of people from across the globe interact; hence, it is an excellent platform for the business world. Generally, blockchain is the newly trending technology used for the purpose of verifying as well as storing of transaction records for online cryptocurrencies such as Bitcoin (Getso & Johari, 2017). According to Holotescu (2018), blockchain technology can be used in the creation of a decentralized environment that prohibits any third-party organization from controlling the cryptographically validated transactions and data. It records any transaction that has been completed in a ledger, to ensure that it can neither be changed nor invalidated. Similarly, Zheng, Xie, Dai, Chen, and Wang (2017) refer to blockchain as a public ledger where all transaction that has been completed are stored in a list of blocks.

A term that is closely associated with blockchain technology is the Cryptocurrency can be defined as a digital asset built to facilitate completed transactions using cryptography. It helps in providing protection to the completed transactions and controlling the creation of additional units of the currency (Okhuese, 2017). Cryptocurrency benefits significantly from the blockchain technology. For example, blockchain technology users utilize public and private keys to develop digital signatures as well as transactions within the cryptography in a secure way (Yaga, Robi, & Scarfone, 2018). Similarly, blockchain network which utilizes mining has the capacity of deciphering intricate puzzles when using cryptographic hash functions (Yaga et al., 2018). Therefore, cryptocurrency is strengthened by the blockchain technology due to its heavy utilization of cryptographic services.

The implementation of the blockchain technology allows the development of new models of economies and marketplaces, governance, identification, and more. Blockchain infrastructure can be used to adopt decentralized sharing economy business models as an alternative to the centralized business models by eliminating the intermediaries and substitute them with a peer-to-peer network, making it possible to carry out direct transactions between two parties (Tumasjan & Beutel, n.d). Additionally, blockchain economy can appear in a new form known as decentralized autonomous organization characterized by rules of governance specified in the blockchain. As a result, completed transactions would be enforced autonomously with strict adherence to rules defined by intricate contracts (Beck, Muller-Bloch, & King, 2018). The new models of business make governance easier than before since no third party will be consulted in order to identify customers and access market places.

BACKGROUND

The existence of cryptocurrencies is based on blockchain technology. Blockchain technology refers to a distributed ledger that records completed transactions through digital means in a way that cannot be mutated (Shah & Jani, 2018). On the other hand, cryptocurrency is a digital asset designed for the purpose of acting as a medium of exchange by applying cryptography to not only provide protection to the completed transactions but also regulate the production of more cryptocurrency (Okhuese, 2017). The two technologies in conjunction can be implemented in different sectors such as the government, education, banking, healthcare, finance, and private sector. Governments benefit from the application of blockchain technology in numerous ways including in the issuance electronic identity cards (e-ID) to its citizens, developing electronic voting systems, managing land registers, recording and sharing medical information (Jun, 2018). These benefits are an evidence that there are numerous blockchain projects in many countries. As of 2017, majority of powerful actors on global scene have either planned, announced, or integrated the adoption of blockchain technology Figure 1.

The concept of blockchain technology has been applied extensively in the field of education for various purposes. First, to issue certificates known as blockcerts, which have four essential components: issuer, certificates, verifier, and wallet (Bartolome, Bellver, Castaneda, & Adell, 2017). Second, schools use blockchain technology to curb degree fraud. In this case, the users from different parts of the world can check if the student's ID are in alignment with the data stored in blockchain to confirm the credibility of the degree (Chen, Xu, Lu, & Chen, 2018). Lastly, blockchain technology can be used to store a verified e-portfolio and manage intellectual property in schools (Grech & Camilleri, 2017). In the banking sector, blockchain has been adopted through electronic payments to improve the banking system (Shah & Jani, 2018). Furthermore, blockchain technology is used to solve various issues that banks encounter such as confusing possession of data user, limited and poor quality of data, problems associated with data sharing between institutions (Guo & Liang, 2016). In the contemporary environment, banks rarely encounter such challenges due to the introduction of the blockchain technology.

Figure 1. Region wise global impact of blockchain technology in public sector

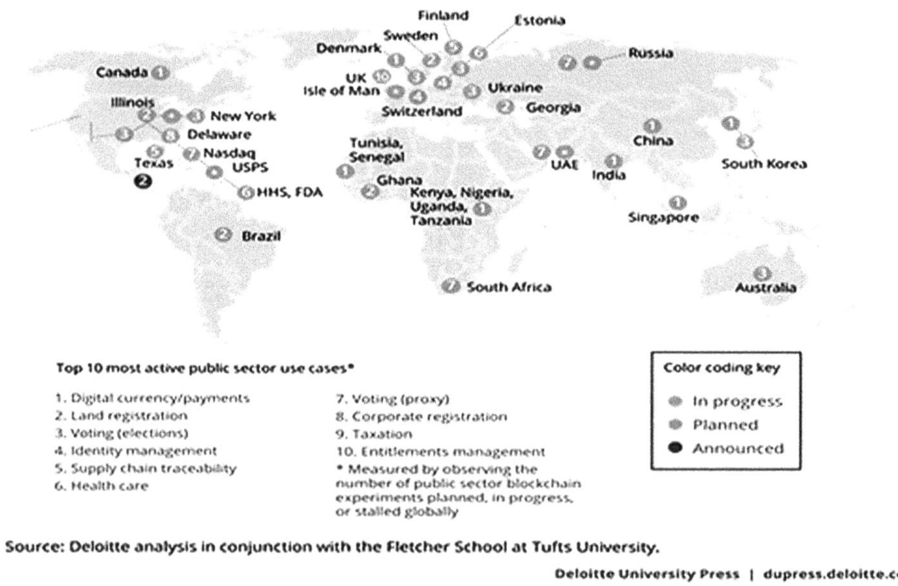

**For a more accurate representation see the electronic version.*

According to Gordon and Catalini (2018), blockchain technology can be used in the healthcare sector to access guidelines through digital platforms, aggregate and liquidate data, determine identity of patients, and to ensure data are immutable. Application of blockchain technology has made it possible for hospitals to solve problems related to patient-driven interoperability involving governance, security issues, and privacy. Heston (2017) reiterates that this greater interoperability is significant to the healthcare providers since it aids in the improvement of how they access their medical records, take photographs of their archives, and prescribe where they store data. Finally, the use of blockchain technology in the financial system assists in mitigating the costs such as the procedure of settlement along with compliance teams incurred by customers when charged by the banks (Casey, Crane, Gensler, Johnson & Narula, 2018). According to Casey et al. (2018), the working of blockchain technology combined with crypto ecosystem leads to crowdfunding along with new secondary marketing models of trading on crypto-exchanges. It is now apparent that government, education, banking, healthcare, finance, and private sectors have benefited immensely through the use of blockchain and cryptocurrencies. Figure 2 reveals that blockchain adoption is increasing across multiple sectors.

Figure 2. Global Blockchain Adoption Phase by Industry
Source: Statista, 2018

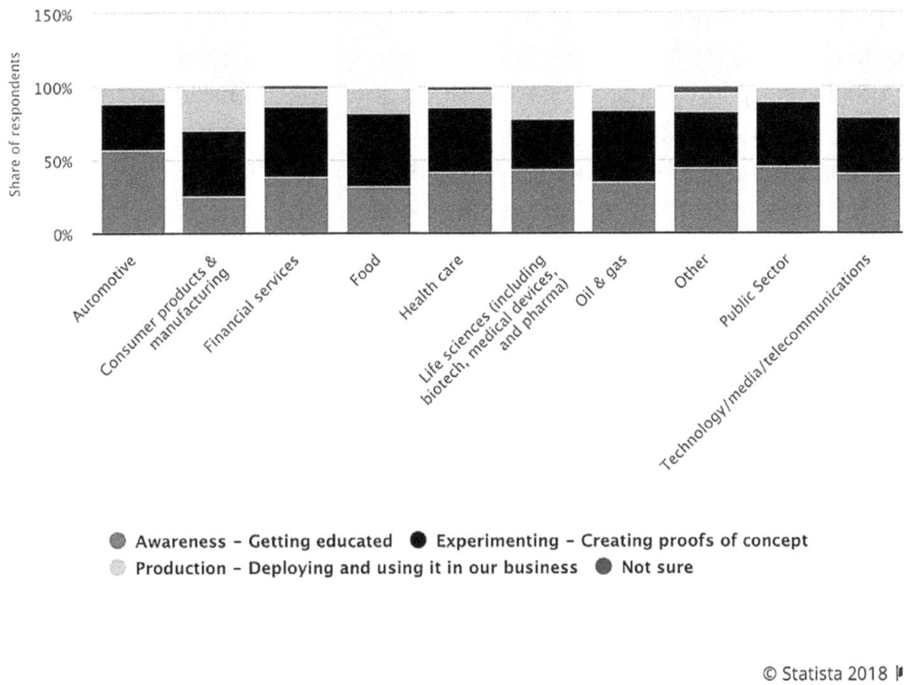

**For a more accurate representation see the electronic version.*

BLOCKCHAIN TECHNOLOGY AND CRYPTOCURRENCIES SIGNIFICANCE

The contemporary society is a product of transition stages that have taken place from industry-based society to a digital one. In the digital society, almost all interactions are carried out virtually by utilizing the present digital technologies such as mobile phones (GSMA, 2015). As such, those living in a digital society have the ability to interact with both public as well as private sectors including education, utilities, transportation, and healthcare, regardless of time and place, resulting in increased efficiency and productivity (GSMA, 2015). Therefore, in a digital society there no central point of control, also known as the third party.

Conversely, in industry-based society interactions between public and private sectors depend on the central point of control. The importance of blockchain technology and cryptocurrency in the modern world transition stages from industry-based society to a digital society is eliminating the central point of control (third party) and alternatively create a peer-to-peer network which permits direct transactions between two parties (Tumasjan & Beutel, n.d). An example of digital society is mobile banking which has eliminated the third party at the bank counter. Today, blockchain technology cryptocurrency has made it possible to access your bank account without visiting the bank counter.

Technological advancements have drastically changed the footprint of numerous industries in the recent decades. In this regard, blockchain technology and cryptocurrencies are the next technological breakthrough which will change functions of these industries. According to Hreinsson (2018), blockchain technology has mostly been used to develop efficiencies, which could eventually make it become one of the most remarkable innovations in the modern world. As a result, it can be predicted that the blockchain technology has immense potential and vast possibilities in future (Hreinsson, 2018). Similarly, the prominence of cryptocurrencies relies on the importance of developing its underlying blockchain (Hreinsson, 2018). Therefore, both blockchain and cryptocurrencies have the potential to improve the efficiency of many industries in the short-term.

Blockchain technology and cryptocurrencies can be used to enhance the economic development of a country in different ways. First, it can facilitate quick and cheaper international development. It is important to note that the cost and inefficiency related to international payment across specific corridors has the ability to impede economic growth (Pisa & Juden, 2017). Second, it will ensure transparency by reducing fraudulent activities at public domain based on the fact that blockchain technology prohibits any central authority from modifying records of the completed transaction (Ahmed, 2017). This is an economic practice that helps a country to prevent financial loss from public coffers that have been allocated for economic development.

The government can also use blockchain technology and cryptocurrencies to reduce the economic and political costs of environmental governance. According to Catalini (2017), crypto-tokens and blockchain can be used to reduce the cost of networking and the expenses incurred when verifying the completed transactions. In today's environment, the government can use blockchain technology to conduct costless verification, reduce the cost of conducting financial audits of completed transactions, and permit the emergence of new marketplaces (Catalini & Gans, 2016). As a result, markets can facilitate the exchange of goods and services between buyers and sellers without relying on the traditional trusted third parties, thus reducing networking cost.

LITERATURE REVIEW

The Technological Foundation of Blockchain Technology

Operation of blockchain technology is based on the peer-to-peer model which eliminates the need for intermediaries, also known as "trusted third parties." Crosby, Nachiappan, Pattanayak, Verma, and Kalyananaraman (2015) describe the operation of blockchain technology during a financial transaction from person A to person B in six stages as illustrated in fig 3. In the first stage, person A wants to send money to the recipient. In the second stage, the transaction is represented online in form of a "block." In the third stage, each part of the network receives broadcast of the block. In the fourth stage, people in the network give approval of the transaction to determine its validity. In the fifth stage, the block can be attached to chain in a way that it provides transparent record of transaction that cannot be deleted. In the final stage of blockchain's operation, the money is transferred from person A to the recipient. Comprehension of the blockchain technology operation requires one to know the components of the blockchain. According to Bruyn (2017), blockchain technology has eight components namely: database, block, hash, miner or node, transaction, fork, safety, and incentives.

Operation of blockchain technology depends on its three security features. The first feature is the ledger which records each transaction completed. The immutable nature of the ledger renders the existing data indelible. The ledger is a decentralized application prohibiting any third party from accessing any sensitive data it stores (Stephen & Alex, 2018). In other words, the ledger records its data in a read-only form. The second security feature is the chain of block, each of which is attached to the previous hash. Every block has a hash which corrects any changes made on the data by an attacker (Stephen & Alex, 2018). In short, the block is arranged in a chain to ensure the sensitive data is secure. The third security feature of blockchain technology is its decentralized application which facilitates successful peer-to-peer communication (Stephen & Alex, 2018). All the three features protect blockchain technology from fraud transactions.

The inception of the blockchain technology has improved security in different ways. First, blockchain technology uses the proof-of-work and proof-of-state mechanisms in determining if the data is secure and reduce losses stemming from fraudulent activities (Lia, Jianga, Chenb, Luoa, & Wen, 2018). Additionally, it enhances the security through a two-factor authentication

Figure 3. How does blockchain work?
Source: Copley, 2017

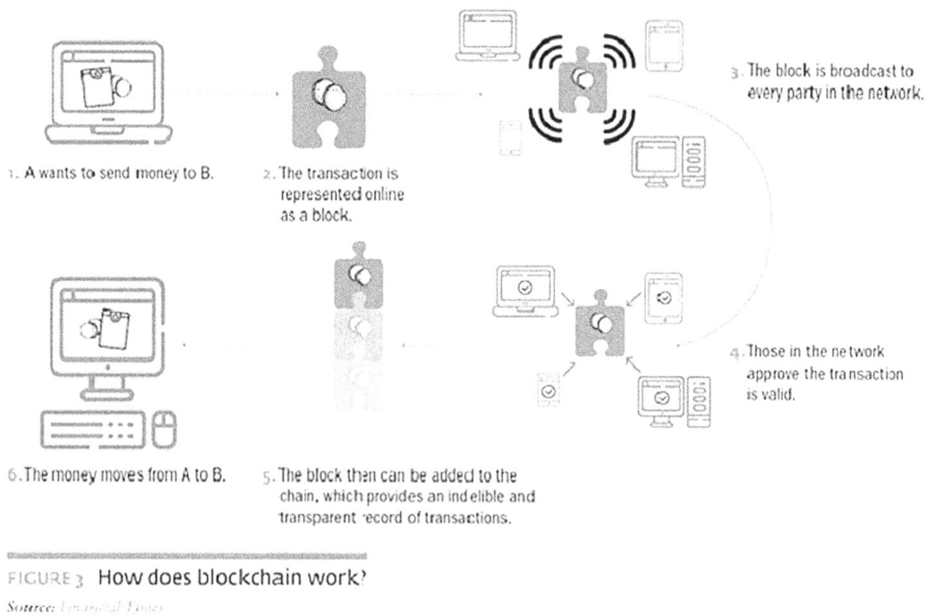

which has been considered as the optimal method to strengthen authentication (Park & Park, 2017). Another security measure is the implementation of Hashed Timelock Contracts with bi-directional payment channels which permits payment to be routed across several peer-to-peer payment channels in a secure way (Lin & Liao, 2017). Although blockchain technology has made tremendous efforts to improve upon its security issues, it stills encounters challenges.

The challenges blockchain technology encounters are associated with security of transactions. The first challenge is known as double-spending security threats which results from a mechanism of verifying transactions. It is an attack emerging if a consumer utilizes the same cryptocurrency repeatedly during transactions (Mosakheil, 2018). For example, a merchant would exhaust all money and goods if a customer creates a longer tail containing a reverse transaction using similar input reference. The second security issue blockchain technology encounters is the 51% attack which provides hackers with the authority to destabilize the entire network of blockchain. In this case, the hackers will be able to prevent all or specific valid block from being mined and reverse transaction in their favor as a result, block will be created faster as compared to others in the network and the private fork gets longer as compared to the primary bitcoin fork the genuine network miners built (Mosakheil, 2018). Another security issue affecting blockchain technology is selfish mining attack characterized by failure of a hacker to publish as well as disseminate a valid solution to other miners in the network. In such a scenario, the hacker chooses the block to be released or publishes numerous blocks concurrently and thus compelling other miners in the network to remove their block and incur losses (Mosakheil, 2018). It is worth mentioning that the attacker hardly acts like a regular miner.

Blockchain as Digital Finance Platform

As mentioned before, the existence of cryptocurrency depends on the blockchain technology due to the significant use of cryptography. Reo (2014) defines cryptocurrency as a virtual medium of exchange whose existence is based on ledger entries in the peer-to-peer network. According to DeVries (2016), cryptocurrency is a technology which was developed a decade ago for the purpose of facilitating digital trade by providing a secure, encrypted, peer-to-peer network. Operation of cryptocurrency is based on the theory of providing the solution to encryption algorithms to create finite unique hashes. Users are able to exchange hashes the same way they can exchange physical currency when cryptocurrency is combined with a system of computers that verifies transactions, (DeVries, 2016).

Byzantine General Problem and the double-spending problem has been the major problems solved by bitcoins using cryptocurrency. The Byzantine General Problem refers to a longstanding problem in the computer science literature where distributed parties find it difficult to reach consensus because they lack trust in each other (Dourado & Brito, 2014). The double-spending problem stems from employing a trusted third-party intermediary to verify the completed transaction (Dourado & Brito, 2014). To solve the two problems, cryptocurrency has been used to eliminate the intermediary who might not be trusted and also reduce the cost of hiring the third party.

Cryptocurrency benefits from the utilization of the blockchain technology as it helps in reducing the challenges arising from Byzantine General Problem and the double-spending problems. Since the solution to these issues require elimination of the third parties, blockchain technology intervenes by creating a decentralized environment. According to Yli-Huumo, Ko, Choi, Park, and Smolander (2016), the main reason for the interest in blockchain is its ability to provide secure transactions, anonymity alongside data integrity without including any third-party organization to monitor transactions. Therefore, it is evident that the efficiency of cryptocurrency is determined by blockchain technology which creates a decentralized environment to eliminate the intermediaries.

Due to the versatile nature of these technologies, both of them can be applied to various sectors like government, education, healthcare, finance, among others. In the financial sector, they can be used to increase the efficiency in providing post-trade services such as collateral management, stock lending, and settlement custody (Gantori et al., 2017). Blockchain technology and cryptocurrencies can be used to improve governance. The government can use blockchain technology to adopt a decentralized governance through decentralized autonomous organizations, or societies, and the creation of the decentralized system of voting that resists top-level domain censorship (Atzori, 2015). According to Narayanan, Bonneau, Felten, Miller, and Goldfeder (2016), blockchain governance is concerned with determining who has the authority among the internal and external actors, in what form, and at which level (as cited in Hsieh, Vergne, & Wang, 2017). By decentralization, the blockchain technology controls who has the authority to determine validity of completed transactions at any form and level. Nagarajan (2018) maintains that the structure of cryptocurrency is inherently decentralized and argues that its implementation relies on the proper mutual interaction between internal and external actors forming the basis of non-technical governance structures.

In the field of education, blockchain technology can be utilized in different ways. First, it can be used for an automatic transfer of credits when a student changes the learning institution (Briggs, 2018). Second, the professor can use blockchain technology to verify exam results to confirm if the student has completed obligations of the course before storing the information in the centralized database. After that, the professor or the school administration will be able to access blockchain address of the student in the central database (Turkanović, Hölbl, Košič, Heričko, & Kamišalić, 2018). Another application of blockchain technology in education is connecting learning institutions in a web of trust to provide not only credibility for each institution but also digital curriculum vitae of their alumni (Kuvshinov, Nikiforov, Mostovoy, Mukhutdinov, Andreev, & Podtelkin, 2018). According to Briggs (2018), other applications of blockchain technology in the field of education include copyright for education content, multi-step accreditation, payment of tuition and application fees, online learning support, interactive learning and analytics, ownership of learning by students, verifying identity and securing information, storing permanent records, and augmenting the current model of education.

Application of blockchain can be used to reduce healthcare complexities like storing medical records. Today, implementation of blockchain technology has enabled the healthcare sector to reduce the cost of medication through better insurance claims and improve the data auditing through the permanent records the blockchain maintains (Heston, 2017). Additionally, with the help of miners and blockchain system, cryptos have the ability to facilitate the transaction between different parties, involving greater transparency in payment of medical bills, secure data sharing across platforms, and mitigating impediments to entry into the healthcare system (Omnicom Health Group, 2018). In the financial sector, blockchain technology has made digital currencies possible resulting in the era of online banking, payment, and transactions, all of which hardly depend on physical currencies (Lewis, McPartland, & Ranjan, 2017). According to Miro (2016), blockchain technology has enabled the financial institution to access all the information required in real time to address the needs of their customers directly. As a result, customers are getting direct services without involving the intermediaries who might delay the banking services.

The rising incidences reported on surveillance and security breaches compromising the privacy of users, has raised the need to secure and hide the confidential data of various companies around the world including banks, among other money transfer companies. To implement this, a protocol that turns a blockchain into an automated access-controls manager excludes the need to trust a third party (Zyskind, Nathan, & Pentland, n.d). Eliminating the intermediary plays a vital role in protecting many companies around the world from people whole might have ill intentions. The blockchain technology can be used to create a decentralized, peer-to-peer networking prohibiting one central server from accessing and processing data belonging to all companies in the network (Batlin, Jaffrey, Murphy, Przewloka, & Williams, 2016). Creating such a network can result in the innovation of a financial system characterized by solitary universal, honest, and completely immutable ledger (Batlin et al., 2016). Therefore, the creation of a decentralized, peer-to-peer networking helps to conceal as well as provide protection to many companies around the world including banks, among other money transfer companies from the control of an intermediary. Projections of blockchain business value indicate high ratio of technology's integration in financial and corporate sectors Figure 4.

Figure 4. Blockchain business value forecast 2018 – 2030

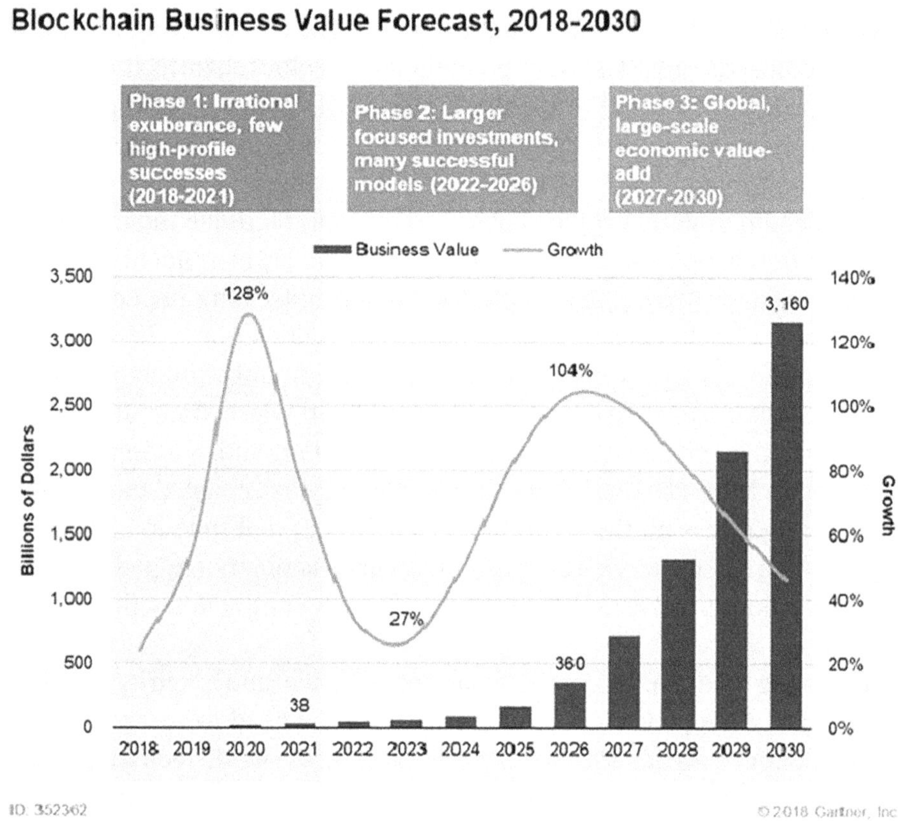

Governance and Legal Framework in Blockchain Technology

Over recent years, the application of the blockchain technology has been associated with governance. Without referring to blockchain, it is important to assess how governance operate within major institutions along with the current internet to assist in providing decentralized governance to certain level. The concept of governance has raised a debate usually led by influential institutions like the federal government (Curran, 2018). There are two categories of blockchain governance: the off-chain governance which nearly like the traditional governing structures, and the on-chain governance which embodies the modern iteration of governance in blockchains and it is accompanied by interesting and polarizing concepts (Curran, 2018).

Blockchain technology is something that can be applied in various fields associated with governance.

Blockchain governance helps to record information in a manner that can, later on, be verified as authoritative. In this case, a blockchain distributes the data it stores; thus, making it easier to access and conversely, making it difficult to eradicate the data entirely. As a result, any individual has the authority to verify on their own that particular entry on a blockchain has not been modified because it was created as well as prove that it was created through a particular process (Stark, 2016). The second importance of blockchain governance is that it can be used to create systems of direct democracy. The best way to make democracy more effective is through direct involvement of citizen in the process of decision making. Such a new model democracy involving public participation can be implemented by blockchain technology (as cited in Atzori, 2017). As a result, the government will be able to practice democracy through consulting its citizen during decision making process.

The third importance of blockchain governance is its ability to create an automated process making it easy to develop a substantially faster and more efficient government system. According to Jun (2018), automating government administrative system using Smart Contracts can lead to a direct governance system. He advices people not to be dreadful of the automating administrative systems of the government because management of the laws implemented in the blockchain requires consents of all members of the community. The fourth significance of blockchain governance is its ability to develop globalized or borderless government services. It is worth mentioning that services based on blockchain can go beyond the geographical limits of a country and elevate their jurisdiction to a worldwide cloud (as cited in Atzori, 2017). Putting into consideration the increasing movement of people between borders, it is recommendable to develop a single overall provider of government services on the basis of multiple nation states.

The fifth importance of blockchain governance is its ability to set ex-ante rule which pre-programs all the conversations through blockchain platform. The moment an event has taken place or specific conditions have been achieved, the transactions are carried out as determined by the programming in the code with the exclusion of intermediaries such as lawyers, judges, or financiers. Since the code in indelible, nobody can modify it and its users are fully aware of the outcomes (Vanderzeil & Currier, 2017). The sixth significance of blockchain technology is that it can be used to create radical transparency that enables all miners in the network to observe each completed transaction. Blockchain technology as a distributed ledger has three unique feature: it records all completed transactions on a list that is duplicated in computers across the world, it continuously and automatically compares all completed transactions against the numerous copies to prevent fraudulent activities, and it is publicly and instantaneously available to all miners in the network, making it mutable to the public, instead of being mutable to a central institution (Vanderzeil & Currier, 2017). It is these features that enables blockchain governance to create a radical transparency.

The seventh importance of blockchain governance is that it can be used to improve liquidity, the ability to transact a considerable amount of a security at a low cost within a short duration. Blockchain governance helps to improve liquidity because it has the ability to not only minimize cost but also reduce the duration needed for conducting and accomplishing secure transactions. This can be achieved regardless of whether blockchain is utilized as the primary platform for sharing registration and exchange, or whether the stock marketers introduce them in a more limited manner for the purpose of streamlining the post-trade along with settlement process (Yermack, 2017). The eighth importance of blockchain governance is that it can be used in the establishment of digital identities for businesses, citizens, among other things associated with government. For example, blockchain technology can be used to manage marriage certificates, death records, birth certificates, passports as well as visa information (Berryhill, Bourgery, & Hanson, 2018). Blockchain identity plays a critical role in governance to determine validity of the identified item and thus preventing fraud.

Finally, blockchain governance helps in creating a credible voting system and prevent fraudulent activities that would result in rigged elections. In most cases, the people who cast their votes are not aware of what happens to their ballots. Fortunately, inception of blockchain technology would make it possible for the voter to not only observe their ballots but also affirm the selections they made. On the other hand, the government will be in position to keep in check the entire system better as well as save money that would otherwise been spent to hire manpower and resources (Medium, 2018). It is important to note that the voting process need huge human labor to use the infrastructure, monitoring, and counting the ballots. Technical glitches in the voting systems has compelled many people to recommend blockchain technology because it assures a transparency that never existed before.

Despite the successful efforts witnessed in many countries leading in the implementation of blockchain technology, some nations have discouraged implementation of cryptocurrencies due to security reasons associated. For example, the technological giant of China, Baidu removed the option Bitcoin as a payment method for security reasons. The same applies to South Korea, which has already ceased the offering of crypto coins, making it to account for only 20 percent of all Bitcoin trades globally (Partanen, 2018). On the contrary, the European Union has succeeded in setting things in motion concerning the implementation of cryptocurrency in the legislature. This could be attributed to Europe's positive reception of this technology and the need for a prompt response to evade bureaucratic problems (Partanen, 2018). Unlike China, the European Union has not banned or limited the operation associated with mining leading to a growing market that establishes a long-term production for the cryptocurrencies. The positive reception of the European Union shows that another major role that can be played by the government to implement cryptocurrencies is through formulating laws advocating for their use.

FUTURE WORK

Much has been discussed about the utility of blockchain technology and cryptocurrencies, the next step entails contemplating about its future implementations. One of the most significant steps for those who are beneficiaries of blockchain technology is to seek influence from both elected and government officials to formulate necessary laws that will allow business as well as individuals to take full advantage of the new technology (Reuters, 2016). Another implication of the applications of blockchain technology in the future is its ability to simplify work and minimize fraud. It worth noting that blockchain technology can mitigate or eliminate manual efforts needed to reconcile and resolve disputes (Chartered Accountants, 2017). Such reconciliations can be used by individuals to increases efficiency in their future business operations. In other words, the utilization of the blockchain technology will make it harder for people to engage in fraudulent activities.

To eliminate malpractices such rigging elections which is a common phenomenon in many nations, application of blockchain technology would be necessary for the near future. As such, one of the government's expectations for blockchain technology and cryptocurrencies implementation will be a transparent election process. In this case, a voter will be able to confirm if the vote cast has been successfully transmitted immediately after voting (Akram, 2017). In short, the government should consider eliminating the manual voting system which is still practiced in many countries. Finally, the ability of blockchain technology to create a decentralized, peer-to-peer environment is expected to be extensively used after its implementation to eliminate flaws and data leakages stemming from reliant on the third party (Train & Levin, 2017). Removal of the third party is crucial in forming a blockchain governance, which is going to be a substantially faster and more efficient government system.

CONCLUSION

The technology of blockchain and cryptocurrencies have made it possible to create a virtual environment where billions of people from across the globe can interact. Blockchain technology refers to a distributed ledger that digitally records completed transactions through digital means. On the other hand, cryptocurrency is a digital asset designed for the purpose of acting as a medium of exchange by applying cryptography to not only protect the completed transactions but also regulate its production. The implementation of the blockchain technology allows the development of new models for economies and marketplaces, governance, identification, and more. Blockchain technology is the basis of cryptocurrencies due to its heavy utilization of cryptographic functions. It can be used to adopt decentralized sharing economy business models as an alternative to centralized business models by eliminating the intermediaries and creating a peer-to-peer network which allows interested parties to transact directly, as in the case of Bitcoins.

Understanding the operation of blockchain technology requires one to comprehend the components of blockchain which includes the database, the block, hash, miner, transaction, fork, and incentives. The inception of blockchain technology has improved security through proof-of-work, proof-of-state mechanisms, two-factor authentication, and the implementation of Hashed Timelock Contracts. Bitcoin is the most popular example of cryptocurrency, which has also solved two major problems in computer science: The Byzantine General Problem and the double-spending problem. Both technologies can be utilized in various fields such as government, education, healthcare, and finance. The rising incidences reported on surveillance and security breaches compromising the privacy of users depict the need to conceal and provide security to the data of companies around the world including banks, among other money transfer companies to protect them from third parties. Over recent years, the application of blockchain technology has been associated with governance. Blockchain governance has helped to to record information in a manner that can, later on, be verified as authoritative, create systems of direct democracy, to create an automated process facilitating effective government system, and to develop globalized or borderless government services.

Governments have played two major roles in blockchain technology and cryptocurrencies implementation: initiating blockchain projects and formulating laws advocating for the use of cryptocurrencies. Despite the successful efforts witnessed in many countries leading in the application of blockchain technology, some countries like China and South Korea have discouraged the use of cryptocurrencies due to security reasons associated. Continued application of blockchain technology and cryptocurrencies requires elected and government officials to formulate necessary laws that will allow business as well as individuals to take full advantage of the new technology. The final implication of blockchain technology will soon alert the voter if the casted vote has been successfully transmitted immediately after voting. To summarize, the inception of blockchain technology and cryptocurrencies has played a critical role in creating a decentralized, peer-to-peer environment which eliminates losses that might arise due to using thirty parties.

REFERENCES

Ahmed, O. (2017). *Block chain technology: concept of digital economics.* Retrieved from https://mpra.ub.uni-muenchen.de/80967/1/MPRA_paper_80967.pdf

Akram, W. (2017). Blockchain technology: Challenges and future prospects. *International Journal of Advanced Research in Computer Science*, *8*(9), 642–644. doi:10.26483/ijarcs.v8i9.4950

Atzori, M. (2015). *Blockchain technology and decentralized governance: is the state still necessary?* Retrieved from http://nzz-files-prod.s3-website-eu-west-1.amazonaws.com/files/9/3/1/blockchain+Is+the+State+Still+Necessary_1.18689931.pdf

Atzori, M. (2017). Blockchain technology and decentralized governance: Is the state still necessary? *Journal of Governance and Regulation*, *6*(1), 45–62. doi:10.22495/jgr_v6_i1_p5

Bartolome, R.A., Bellver, C., Castaneda, L., & Adell, J. (2017). Blockchain in education: introduction and critical review of the state of the art. *Electronic Journal of Educational Technology, 61*, 1-14. Doi:10.21556/edutec.2017.61

Batlin, A., Jaffrey, H., Murphy, C., Przewloka, A., & Williams, S. (2016). *Building the trust engine. How the blockchain could transform finance (and the world)*. UBS AG.

Beck, R., Muller-Bloch, C., & King, L. C. (2018). Governance in the blockchain economy: A framework and research agenda. *Journal of the Association for Information Systems*, 1–40. Retrieved from https://www.researchgate.net/publication/323689461_Governance_in_the_Blockchain_Economy_A_Framework_and_Research_Agenda

Berryhill, J., Bourgery, T., & Hanson, A. (2018). *Blockchains Unchained: Blockchain Technology and its Use in the Public Sector*. OECD Working Papers on Public Governance, No. 28. OECD Publishing. doi:10.1787/3c32c429-en

Briggs, S. (2018). *Blockchain Technology: Can It Change Education?* Retrieved 7th November, 2018 from https://www.opencolleges.edu.au/informed/edtech-integration/blockchain-technology-education/

Bruyn, S. A. (2017). *Blockchain an introduction*. Retrieved from https://beta.vu.nl/nl/Images/werkstuk-bruyn_tcm235-862258.pdf

Casey, M., Crane, J., Gensler, G., Johnson, S., & Narula, N. (2018). The impact of blockchain technology on finance: a catalyst for change. International Center for Monetary and Banking Studies 2.

Catalini, C. (2017). *How blockchain technology will impact the digital economy*. Retrieved from http://ide.mit.edu/sites/default/files/publications/IDE%20Research%20Paper_v0517.pdf

Catalini, C., & Gans, S. J. (2016). Some simple economics of the blockchain. National Bureau of Economic Research. doi:10.3386/w22952

Chartered Accountants. (2017). *The future of blockchain: Applications and implications of distributed ledger technology*. Chartered Accountants Australia and New Zealand.

Chen, G., Xu, B., Lu, M., & Chen, N. (2018). Exploring blockchain technology and its potential applications for education. *Smart Learning Environments, 5*(1), 1-10. doi doi:10.118640561-017-0050-x

Copley, A. (2017). Figures of the week: Blockchain opportunities and challenges in Africa. *Brookings*. Retrieved from https://www.brookings.edu/blog/africa-in-focus/2017/12/15/figures-of-the-week-blockchain-opportunities-and-challenges-in-africa/

Crosby, M., Nachiappan, Pattanayak, P., Verma, S., & Kalyananaraman, V. (2015). Blockchain technology. Sutardja Center for Entrepreneurship & Technology.

Curran, B. (2018). *What is Blockchain Governance? Complete Beginner's Guide*. Retrieved November 19, 2018 from https://blockonomi.com/blockchain-governance/

DeVries, D. P. (2016). An analysis of cryptocurrency, bitcoin, and the future. *International Journal of Business Management and Commerce, 1*(2), 1–9.

Dourado, E., & Brito, J. (2014). The New Palgrave Dictionary of Economic Cryptocurrency. doi:10.1057/978-1-349-95121-5_2895-1

Gantori, S., Donovan, P., Ganesh, K., DeMichiel, M., Dennean, Trussardi, F., & Klien, M. (2017). Cryptocurrencies. *Chief Investment Office Americas, Wealth Management*. Retrieved from https://www.ubs.com/content/dam/WealthManagementAmericas/cio-impact/cryptocurrencies.pdf

Getso, M. M. A., & Johari, Z. (2017). The blockchain revolution and higher education. *International Journal of Information System and Engineering, 5*(1), 57–66. doi:10.24924/ijise/2017.04/v5.iss1/57.65

Gordon, J. W., & Catalini, C. (2018). Blockchain technology for healthcare: Facilitating the transition to patient-driven interoperability. *Computational and Structural Biotechnology Journal, 16*, 224–230. doi:10.1016/j.csbj.2018.06.003 PMID:30069284

Grech, A., & Camilleri, F. A. (2017). Blockchain in education. Joint Research Centre. doi:10.2760/60649

GSMA. (2015). *Building digital societies in Asia*. Retrieved from https://www.gsmaintelligence.com/research/?file=bd5b3cf1d0533f9c9641039ba6966864&download

Guo, Y & Liang, C. (2016). Blockchain application and outlook in the banking industry. *Financial Innovation, 2*(24), 1-12. DOI doi:10.118640854-016-0034-9

Heston, T. (2017). A case study in blockchain health care innovation. *International Journal of Current Research, 9*(11), 60587–60588. doi:10.22541/au.151060471.10755953

Heston, T. (2017). *Why Blockchain Technology Is Important for Healthcare Professionals*. SSRN Electronic Journal. doi:10.2139srn.3006389

Holotescu, C. (2018). *Understanding blockchain technology and how to get*. Paper presented at the 14th International Scientific Conference eLearning and Software for Education, Bucharest, Romania. doi: 10.12753/2066-026X-18-000

Hreinsson, M. E. (2018). *The future of blockchain technology and cryptocurrencies*. Retrieved from https://skemman.is/bitstream/1946/30832/1/The%20future%20of%20blockchain%20technology%20and%20cryptocurrencies.pdf

Hsieh, Y., Vergne, J., & Wang, S. (2017). The internal and external governance of blockchain based organizations: Evidence from cryptocurrencies. In M. Campbell-Verduyn (Ed.), *Bitcoin and Beyond: Blockchain and Global Governance*. RIPE/Routledge Series in Global Political Economy. doi:10.4324/9781315211909-3

Jesse Yli-Huumo, J., Ko, D., Choi, S., Park, S., & Smolander, K. (2016). Where Is Current Research on Blockchain Technology?—A Systematic Review. *PLoS One*, 1–27. doi:10.1371/journal.pone.0163477

Jun, M. (2018). Blockchain government - a next form of infrastructure for the twenty-first century. *Journal of Open Innovation, 4*(7), 2-12. DOI doi:10.118640852-018-0086-3

Kuvshinov, K., Nikiforov, I., Mostovoy, J., Mukhutdinov, D., Andreev, K., & Podtelkin, V. (2018). *Disciplina: Blockchain for Education*. Retrieved from: https://disciplina.io/yellowpaper.pdf

Lewis, R., McPartland, J., & Ranjan, R. (2017). Blockchain and financial market innovation. *Economic Perspectives*, 1–13.

Lia, X., Jianga, P., Chenb, T., Luoa, X., & Wen, Q. (2018). A survey on the security of blockchain systems. *Future Generation Computer Systems*, 1–25. doi:10.1016/j.future.2017.08.020

Lin, L., & Liao, T. (2017). A survey of blockchain security issues and challenges. *International Journal of Network Security, 19*(5), 653-659. DOI: (5).01 doi:10.6633/IJNS.201709.19

Medium. (2018). *Blockchain Technology Governance in Voting and Survey*. Retrieved November 19, 2018 from https://medium.com/@feelium/blockchain-technology-governance-in-voting-and-survey-91c162122556

Miro, G. J. (2016). Strategic innovation in financial sector: Blockchain and the case of Spanish banks. KTH Industrial Engineering and Management. *Industrial Management (Des Plaines), SE-100*, 44.

Mosakheil, H. J. (2018). *Security Threats Classification in Blockchains*. Retrieved from https://repository.stcloudstate.edu/cgi/viewcontent.cgi?referer=https://www.google.com/&httpsredir=1&article=1093&context=msia_etds

Nagarajan, M. (2018). An Analysis of Cryptocurrency Governance. *Joseph Wharton Scholars*. Available at https://repository.upenn.edu/joseph_wharton_scholars/51

Okhuese, O. A. (2017). *Introducing cryptocurrency. A project presented on Cryptocurrency - The Future of Digital Money*. Retrieved from https://www.researchgate.net/publication/320616742_INTRODUCING_CRYPTOCURRENCY

Omnicom Health Group. (2018). *Cryptocurrency in Healthcare*. Retrieved from http://www.omnicomhealthgroup.com/pdfs/Cryptocurrency-in-Healthcare.pdf

Park, H. J., & Park, H. J. (2017). Blockchain Security in Cloud Computing: Use Cases, Challenges, and Solutions. *Symmetry Journal, 9*(164), 1-13.

Partanen, C. (2018). *The viability of cryptocurrency in relation to the response of financial institutions and governments*. Retrieved from https://www.theseus.fi/bitstream/handle/10024/152122/Christian_Partanen_Bachelors_Thesis.pdf?sequence=1&isAllowed=y

Pisa, M., & Juden, M. (2017). *Blockchain and Economic Development: Hype vs. Reality. Center for Global Development.*

Reo, T. (2014). *How Do Cryptocurrencies Work?* Paper presented at the Inaugural CAIA-SKBI Cryptocurrency Conference, Singapore Management University.

Reuters, T. (2016). *Conclusion and Thoughts about the Future*. Retrieved from https://www.hklaw.com/files/uploads/Documents/Press%20Releases/Blockchain_Ch_9_UPDATED.PDF

Shah, T., & Jani, S. (2018). *Applications of blockchain technology in banking & finance*. Retrieved from https://www.researchgate.net/publication/327230927_Applications_of_Blockchain_Technology_in_Banking_Finance

Stark, J. (2016). *Making sense of blockchain governance applications*. Retrieved 7th November, 2018 from https://www.coindesk.com/making-sense-blockchain-governance-applications/

Statista. (2018). *Blockchain adoption phases in organization worldwide as of April 2018, by industry*. Retrieved from https://www.statista.com/statistics/878748/worldwide-production-phase-blockchain-technology-industry/

Stephen, R., & Alex, A. (2018). A Review on BlockChain Security. *Materials Science and Engineering*, 1–7. doi:10.1088/1757-899X/396/1/012030

Train, T., & Levin, M. (2017). *Blockchain, the future opportunity for trading progression?* Retrieved from http://www.diva-portal.org/smash/get/diva2:1109208/FULLTEXT01.pdf

Tumasjan, A., & Beutel, T. (Forthcoming). Blockchain-based Decentralized Business Models in the Sharing Economy: A Technology Adoption Perspective. In *Business Transformation through Blockchain* (vol. 2). Cham, Switzerland: Palgrave Macmillan.

Turkanović, M., Hölbl, M., Košič, K., Heričko, M., & Kamišalić, A. A. (2018). EduCTX: A Blockchain-Based Higher Education Credit Platform. *IEEE Access: Practical Innovations, Open Solutions*, *6*, 5112–5126. doi:10.1109/ACCESS.2018.2789929

Vanderzeil, S., & Currier, E. (2017). *Governance and the Ungovernable: Implications of Blockchain Proliferation.* Cornerstone Capital Group. Retrieved from https://cornerstonecapinc.com/wp-content/uploads/Governing-Ungovernable_Blockchain-June-2017.pdf

Yermack, D. (2017). Corporate Governance and Blockchains. *Review of Finance*, 7–31. doi:10.1093/rof/rfw074

Zheng, Z., Xie, S., Dai, H., Chen, X., & Wang, H. (2017). *An overview of blockchain technology: architecture, consensus, and future trends.* Paper presented at the IEEE 6th International Congress on Big Data. 10.1109/BigDataCongress.2017.85

Zyskind, G., Nathan, O., & Pentland, A. (n.d.). *Decentralizing Privacy: Using Blockchain to Protect Personal Data.* Retrieved from https://enigma.co/ZNP15.pdf

Chapter 7
Profile Analysis of Research Output in Biotechnology in Turkey

Yaşar Serhat Yaşgül
Marmara University, Turkey

ABSTRACT

The aim of this chapter is to reveal the research profile and intellectual accumulation of Turkey in the field of biotechnology by using bibliometric methods for the period 1973-2016. To this end, the period studied in this framework is divided into two sub-periods (1973-2003 and 2004-2016). Thus, it was aimed to analyze the effects of the Vision 2023 strategy document, which appeared in the 2000s and which considers biotechnology as one of the strategic technologies to be developed as a priority. Using the dataset created from the scanned records in the ISI Web of Science database, Turkey's publication trends, collaboration patterns, preferred journals and research subject areas were identified. Despite the improvement observed in publication performance and cooperation for the 2004-2006 sub-period, Turkey has lagged behind countries such as Brazil, India, and China, especially in terms of publication performance.

DOI: 10.4018/978-1-5225-9012-5.ch007

INTRODUCTION

Participation in new information and information production or existing information and information networks has become one of the basic engines of economic growth in the 21st century. (Evans 2010; Evans 2012) In other words, the primary priority for emerging countries in the 21st century is to participate in new information and information networks as soon as possible and to use these networks effectively.

One of the basic conditions for using new information and information networks effectively is production of scientific publication (research output). The production of scientific publications is of great importance for the learning and adaptation of science/ technology. (Niosi, Hanel and Reid 2012) In this context, it is crucial to determine the future prospects for countries' competitiveness and growth, with particular emphasis on the basic features of scientific activities aimed at new technologies and scientific publications that are natural consequences of these activities. (Yaşgül and Güriş, 2015)

Biotechnology, defined as the manipulation of living organisms or parts of these organisms for production of goods and services, is one of the fundamental technologies that will determine the prosperity and wealth levels of countries in the 21st century. (Bartholomew 1997; Rasmussen, 2010; Wong et al. 2012) Biotechnology can be described as a set of technologies that offer new production methods to the traditional sectors. In other words, biotechnology is not an industry but a platform technology used in many sectors such as healthcare systems, agriculture, veterinary products, environment and mining. (Niosi, Bas and Amador 2013) There are three basic characteristics that distinguish biotechnology from other known technologies. (Bartholomew 1997) These are as follows:

1. Biotechnology is completely dependent upon basic research conducted in various fields such as microbiology, biochemistry, genetics and bioengineering.
2. Biotechnology has application areas in different sectors.
3. The process of bringing together basic researches with fields of application in biotechnology involves various uncertainties. Economic, social and institutional structures play a vital role in reducing these uncertainties.

When the abovementioned three basic characteristics are evaluated together, it is seen that the main source of the development process in biotechnology is the scientific knowledge produced under different disciplines. (Niosi, Hanel and Reid 2012) Therefore, the primary priority in this framework for countries aiming to develop themselves in the field of biotechnology is the realization of biotechnology-based scientific knowledge production in different disciplines. The realization of scientific knowledge production requires intensive co-operation between companies, universities and non-university research institutions. Another characteristic of this network is that it is not only national but also international. (Prange and Kaiser 2004) In this way, co-operation networks in the process of modern scientific knowledge production contribute not only to the quality of the produced knowledge, but also to the dissemination of this knowledge. This is particularly important in terms of science and technology policy. (Frenken, Hölzl and Vor 2005) For developing countries, it is crucial to establish a national biotechnological innovation system, which provides institutional and/or financial support for these networks of co-operation. (Bartholomew 1997; Niosi 2010; Rawat 2008)

The fact that the US and UK-led developed countries are the centers both for the production of scientific knowledge and for the commercialization of this knowledge makes it necessary for other countries to determine the growth strategies for this area. The primary priority in the process of evaluating these framework strategies is the establishment of a scientific publication profile that reflects the production of scientific knowledge. A review of the publication profile will clearly reveal the following:

1. Evaluation of publication performance and quality,
2. The evaluation of the progress of national and international collaborations, which are vital for biotechnology with respect to new technologies, in terms of institution and country,
3. Identify the subject areas of biotechnology publications.

The basic aim of this study is to reveal the research profile and intellectual accumulation of Turkey in the field of biotechnology between 1973 and 2016. In this context, the study period was divided into two subdivisions (1973-2003 and 2004-2016) so that the impact of the Vision 2023 strategy document, which considers biotechnology as one of the strategic technologies that should be developed primarily in the 2000s, can be seen more clearly. However, such a comprehensive study that assesses the efforts of Turkey in this field has not yet been found in literature. However, carrying out such an evaluation is of vital importance in terms of reassessing the strategies for the targets to be achieved and making the needed policy changes. The main motivation behind conducting this study is to make up for this gap in the literature. Biotechnology policies in Turkey will be briefly summarized in the second section. The methodology and database used will be presented in the third section. The database will be analyzed from the perspective of general publishing trend, collaborations and research field structure in the fourth section. The findings will be summarized in the fifth and last section.

BIOTECHNOLOGY IN TURKEY

In this study, it is aimed to investigate the publication profile of Turkey in the field of biotechnology using bibliometric methods. In general terms, it is possible to mention three basic strategic documents in terms of policies to be implemented/implemented in Turkey in the field of biotechnology. The first one was the "Turkish Science Policy: 1983-2003" prepared by the State Ministry in 1983; the second one was the "Turkish Science and Technology Policy: 1993-2003" prepared by TÜBİTAK for the BTYK (Science and Technology Supreme Council) and the third one was the document "Vision 2023: Science and Technology Strategies" prepared in 2001 as the Turkish National Science and Technology Strategy Document for the period 2003-2023. (Erbaş 2008)

Although the importance of biotechnology as a strategic field of technology was realized in the mid-1980s parallel to the developments in the world, a comprehensive action plan for this area was not established in Turkey until 2000s. (Başağa and Çetindamar 2006) In other words, there is a lack of clarity about what kind of policy to pursue in spite of the fact that the basic problem is not very new in terms of policy development related to the issue (Erbaş 2008).

At the sixth BTYK meeting held in 2000, the decision on the preparation of the National Science and Technology Policy Strategy (2003-23) was taken and it was decided that the decision taken in 1999 would be assessed within the scope of the strategy document. It has also been decided at the same meeting to prepare the National Biotechnology Research Program for Agriculture, which would cover the years 2001-10. (Başağa and Çetindamar 2006) Within the scope of Vision 2023 project, biotechnology was considered as one of the 8 strategic technologies to be developed and it was envisaged to focus on four main traditional sectors: healthcare, agriculture, animal husbandry and industry. (TÜBİTAK, 2004) The Turkish Biotechnology Strategy and Action Plan 2015-2018 was announced by the Ministry of Science, Technology and Industry in Turkey in 2015 with a total of 6 targets and 25 relevant action plans. (TC Science Ministry of Technology and Industry 2015)

As of the end of 2013, there are 144 biotechnology firms in Turkey. 56 of these companies operate in healthcare biotechnology, 33 in industrial biotechnology, 23 in agricultural biotechnology and 5 in bioinformatics. In addition, 27 companies are the private suppliers of the sector. (Tunçgenç 2014)

With respect to the trained workforce in the field of biotechnology, the number of undergraduate programs in molecular biology, genetics and bioengineering established at universities reached 58 in 2013. Some of these universities offer master's and doctoral programs. Some of these master's and doctoral programs are conducted under the direction of relevant departments of universities or institutes of science. On the other hand, 18 universities also have research centers or institutes that focus on the field of biotechnology and are organized in an interdisciplinary manner. These universities carry out postgraduate researches and trainings in biotechnology through research centers or institutes. Apart from these, almost all of the medical faculties have specialist training programs in one or both of medical genetic or medical molecular biology branches. (Tunçgenç 2014)

Overall, four critical features have been influential in focusing on Turkey as the subject area of this study.

1. The country has a large population and market potential that can influence the biotechnological development process and enable the commercialization of scientific information to be produced in the biotechnological field,
2. Highly trained manpower in different disciplines focusing primarily on biotechnology,
3. Having traditional sectors such as healthcare, chemistry and food that can be reconstructed on biotechnology,
4. A new vision and its aim to address biotechnology as one of the priority technologies to be developed from the beginning of the 2000s.

It is not possible to think of Turkey's economic transformation independently of the world's innovation agenda. Horizontal technology platforms that have the potential to transform different sectors for rapid transformation in industry offer opportunities. Turkey's technological priorities must be parallel to that of the world. What is critical here is to design which technologies to focus on through which sectors, and the effect of diffusion on other sectors. (Tepav 2015)

METHODOLOGY AND DATA

Bibliometry is an approach that can be used to track the research capacity and knowledge pool of a country. (Martinez, Jaime and Camacho 2014) Bibliometric approaches are used in the field of science and technology to determine the information output of national innovation systems. (Zhao, Li, Han, Li and Li 2016) Bibliographic areas such as author, year and address embedded in publications allow for the assessment of research fields.

Bibliographic studies are also important tools to evaluate co-operation between countries and institutions. In this context, the most effective indicator for measuring collaborations is to take into account joint publications in scientific publications. (Frenken, Hölzl and Vor 2005)

Studies have been carried out previously to assess the scientific basis and research capacity of a country in the field of biotechnology. (Bajwa ve Yaldram 2013; Kumar ve Chand 2005; Molatudi ve Pouris 2006; Gomes Costa et al 2013; Zhao et all 2016; Parakash and Arumugam, 2017; Martinez, Jaime and Camacho 2014;Arencibia-Jorg et al 2016) There are also similar studies involving comparisons between countries. (Frenken et al 2005; Chen and Guam 2011; Glanzel and Zhou, 2011)

Similar to the studies in the literature, the question of how the biotechnological research profile developed in Turkey between 1973 and 2016 was addressed in this sub-period context (1973-2003 and 2004-2016). Thus, the "Vision 2023" document which is expressed as an important strategy plan in the field of science and technology in Turkey has been accepted as an evaluation point. Accordingly, a holistic analysis of the period was made and a comparative evaluation of the sub-periods was presented.

Data Set

The data used in the study were obtained from indexed publications in the subject area of "Biotechnology and Applied Microbiology" at ISI Web of Science, similar to Martinez, et al (2014). Accordingly, a total of 7543 publications were obtained from the SCI Expanded database. Other publications were excluded as Article, Notes, Letters, Review and Proceeding

Papers were included and 6142 publications were included in the study. The address of at least 1 author in these publications is from Turkey. When the 6142 publications included in the review were analyzed in detail, it was found out that the addresses of 9 of the publications were mistaken and not originated from Turkey. For this reason, these 9 publications were excluded from the analysis. As a result, 6133 publications were obtained with at least one author's address from Turkey for the period 1973-2016. Distribution of the publications by periods is shown in Table 1.

The analyzes were carried out in three stages in the study:

1. The general trend of biotechnology researches in Turkey in the period examined (1973-2016) was evaluated.
2. National / international collaborations in the field of biotechnology have been evaluated in terms of partner institutions and countries for the period under consideration.
3. The analyzed database was analyzed comparatively in terms of ISI Subject Areas and the journals published in the third stage.

DATA ANALYSIS

General Publication Trend

The ratio of Turkey's biotechnology publications to the world's biotechnology publications is presented in Figure 1. It is observed that the number of publications have been increasing since 2000s and approximately 1,9% of all biotechnology publications in the world were by Turkish authors as of 2015.

Table 1. Distribution of research outputs in the field of biotechnology in Turkey

	1973-2003	2004-2016	Total
Article	1009	4728	5737
Notes	16	0	16
Letter	10	135	145
Review	10	147	157
Proceeding Paper	41	37	78
Total	1086	5047	6133

Figure 1. Ratio of Turkey's publications in biotechnology to the biotechnology publications in the world

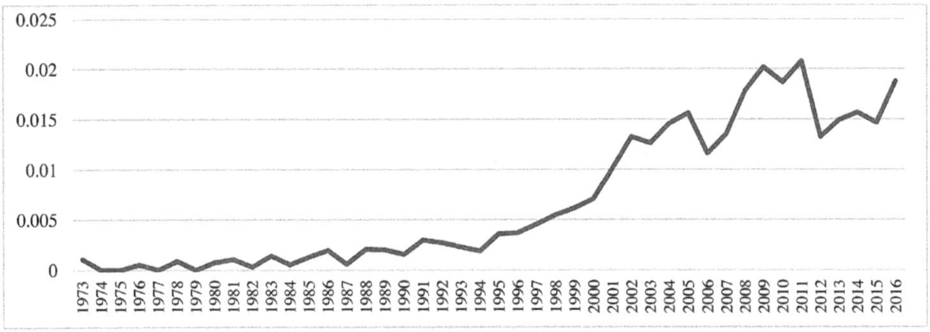

Turkey's publication performance in the field of biotechnology is compared against the performance of emerging countries with strong growth performance in biotechnology such as Brazil, China, India and Argentina in Figure 2, Figure 3 and Figure 4. The global coverage of biotechnology publications is presented comparatively in Figure 2. According to this, China, India and Brazil display a relatively better publication performance. The activity index[1], which sets the level of relative expertise in biotechnology, is presented for Turkey, Brazil, China, India and Argentina in Figure 3. While the activity

Figure 2. Ratio of Turkey's and selected countries biotechnological publications to the rest of the world

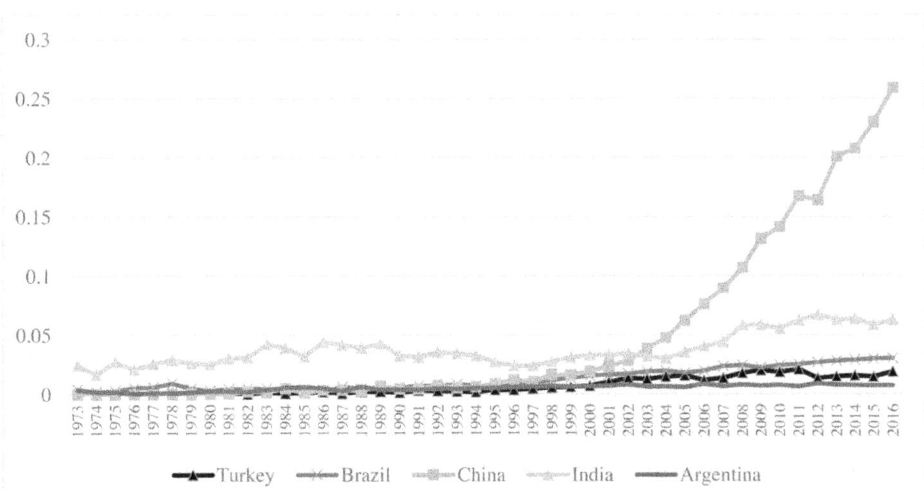

Figure 3. Activity index

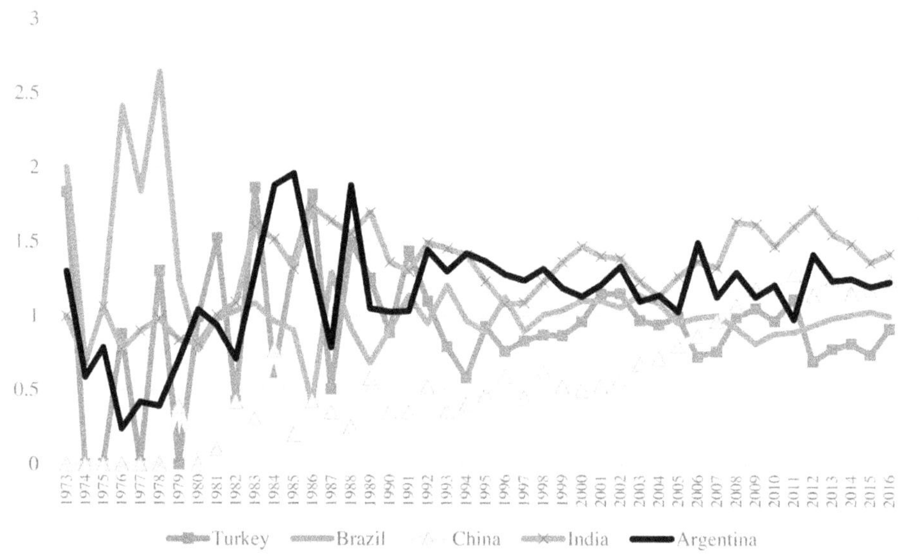

index of the four other countries outside Turkey has revealed a certain level of specialization in the field of biotechnology especially in the last 10 years, Turkey has been following a fluctuating trend in those years and has not revealed the symptoms of a certain specialization in this field.

Figure 4 shows the number of publications in the field of biotechnology per 100 thousand people in those countries. An upward trend is observed in the 2000s in all the countries surveyed. Especially the upward trend displayed by Turkey in 2000s is remarkable. When evaluated on a population basis, the increase in the number of publications in Turkey is more than the other countries examined.

Finally, the average number of citations per publication of the 6133 publications examined in order to evaluate Turkey's overall publication trend in biotechnology has been examined. The 3-year citation window[2] is used to calculate the average number of citations per publication. The average number of citations per publication of the 5178 publications examined for the sub-period of 1973-2014 is 3,30. Figure 5 shows the average number of citations per publication over the years. The average citation rate per publication has increased in 2000s. In fact, the average number of citations per publication for 1086 publications in the sub-period of 1973-2003 was about 2,03, while it

Figure 4. Number of publications on biotechnology per 100,000 people

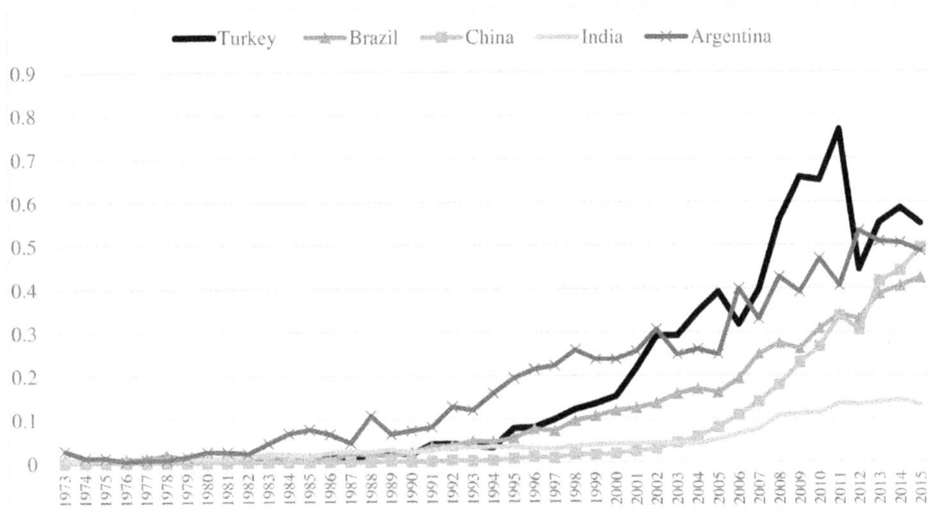

Figure 5. Average number of citations per publication

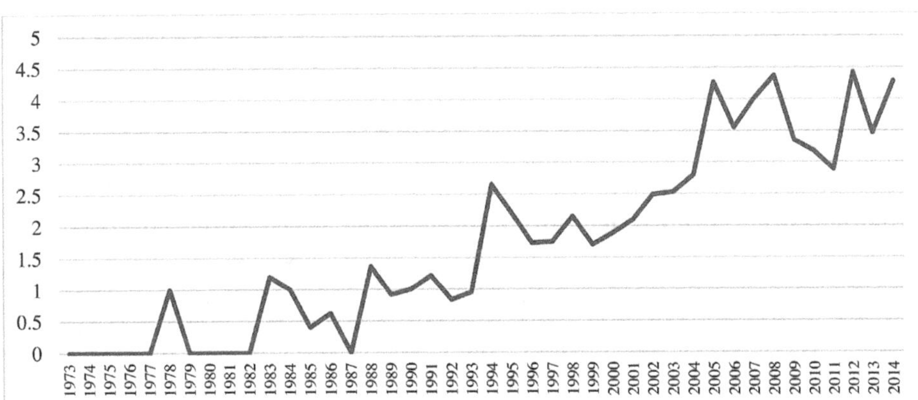

was 3,65 for 4092 publications in the sub-period of 2004-2014. It is possible to say that the contribution to the knowledge accumulation of publications for the sub period of 2004-2014 has increased.

Overall, although Turkey's contribution to the number of publications in the field of biotechnology and the contribution of publications to scientific knowledge accumulation has achieved a certain level of increase, it has not yet reached the expected level. In this context, it is necessary for Turkey to increase the publication performance in biotechnology not only with respect to the activity index indicating a certain level of specialization in the world but also with respect to the ratio of publications worldwide.

Collaboration Patterns

As mentioned in the previous sections of the study, participation in collaboration networks in new technologies such as biotechnology is of great importance for the institutions and countries that determine this technology among the priority technologies to be developed. In this context, it is vital that Turkey's publications on biotechnology are evaluated in the context of collaborations. The 6133 publications were classified under three groups in terms of collaboration similar to Zhao et all 2016. According to this:

1. If all of the author addresses in a publication are from Turkey and include the same institution, it is classified as "publication without collaboration"
2. If the author addresses in a publication are all from Turkey but contain at least two different institutions, it is classified as "publication with national collaboration"
3. If at least one of the author's addresses in a publication contains a non-Turkish address, it is classified as "publication with international collaboration".

The distribution of Turkey's biotechnology publications in terms of collaboration is presented in Table 2. Of the 6133 publications analyzed, 50% consist of publications without collaboration, 31% with national collaboration, and only 19% are with international collaboration. When the situation is analyzed in terms of the sub-periods, it is seen that there is no significant

Table 2. Investigation of all records in the period of 1973-2016 in terms of collaborations

	Total	Number of Publications without Collaboration	Number of Publications with National Collaboration	Number of Publications with International Collaboration
1973-2003	1086	671	209	206
2004-2016	5047	2395	1682	970
Total	**6133**	**3066**	**1891**	**1176**

development in terms of international collaboration. On the contrary, there is a serious increase in studies involving national collaboration. The proportion of studies involving national collaboration for the sub-period 1973-2003 was 19%, compared to 33% for the sub-period 2004-2016. While the development of national collaborations reveals a strong internal capacity development in terms of information dissemination, the lack of adequate development of international collaboration poses a major deficiency in participation in global knowledge networks.

In the period of 1973-2016, a total of 1891 publications based on national collaboration were produced by 498 institutions through a total of 9,109 collaborations [3]. 158 of these 498 institutions are universities. Table 3 lists

Table 3. Distribution of the records of the first twenty national institutions

Institution	Total Number of Records	Number of National Collaborations	Number of International Collaborations	Number of Records that do not Involve Collaboration
Ege University	528	155	89	284
Hacettepe University	490	219	81	190
Ankara University	466	221	78	167
Middle East Technical University	347	112	92	143
Gazi University	315	152	34	129
Istanbul University	315	114	73	128
Atatürk University	257	107	29	121
Dokuz Eylül University	256	60	26	170
Istanbul Technical University	233	79	73	81
Çukurova University	172	73	43	56
Selçuk University	162	64	17	81
Dicle University	160	66	17	77
Boğaziçi University	158	48	54	56
Marmara University	156	75	51	30
Karadeniz Technical University	155	78	19	58
Uludağ University	141	47	21	73
Süleyman Demirel University	138	58	23	57
TÜBİTAK	138	84	27	27
Erciyes University	131	72	22	36

the top 20 national institutions with the highest number of publications in the field of biotechnology between 1973 and 2016 and the distribution of records for these institutions.

As can be seen in Table 3, 19 of the first twenty institutions are public universities. Only TUBITAK, a public research institution, has been able to enter this list. This is quite remarkable because the collaboration between the actors involved in biotechnology is very important in terms of the establishment of scientific knowledge stocks and the flow of knowledge between these knowledge stocks. (Bartholomew, 1997) In this context, the absence of private/public companies and research institutions among the first twenty institutions reveals a considerable lack of knowledge dissemination[4].

When evaluated in terms of the quality of their collaboration, the first 20 national institutions with the highest proportion of international collaborations are Boğaziçi University (34%), Marmara University (33%) and Istanbul Technical University (31%). When the same situation was evaluated in terms of national collaborations, TÜBİTAK (61%), Erciyes University (55%) and Karadeniz Technical University (50%) ranked the highest. The role undertaken by TUBITAK as a public research institution in terms of national collaborations is remarkable. TÜBİTAK has played an important central role in terms of national collaboration network.

Finally, it would be appropriate to examine the territorial distribution of publications with international collaboration. In the period of 1973-2016, Turkey published 1176 publications through 1726 international collaborations with 99 countries. 206 of these publications (244 co-operations with 36 countries) were published in the sub-period of 1973-2003 and 970 publications (1482 co-operations with 99 countries) were published in the sub-period of 2004-2016. Table 4 shows the top 20 countries with the highest level of collaboration between 1973 and 2016.

The share of the first twenty countries in publications with international collaboration is 77%. In this context, it is possible to say that international collaboration is concentrated in these twenty countries in the biotechnology publications of Turkey. Turkey increased its international collaboration 6 times in the 2004-2016 sub-period compared to the 1973-2003 sub-period. As mentioned in previous sections, although the number of publications containing international collaborations is not sufficient, the abovementioned increase is noteworthy. Another remarkable detail concerning Table 4 is that China and India increased collaboration with developing countries showing potential for growth in other areas of biotechnology in the sub-period of 2004-2016. The development of collaboration between countries with similar objectives

Table 4. The number of collaborations of the first 20 countries

Country	Total	1973-2003	2004-2016
USA	391	70	321
UK	161	52	109
Germany	131	15	116
Italy	106	4	102
Netherlands	57	8	49
Japan	55	11	44
France	51	5	46
Spain	51	6	45
Canada	42	3	39
Belgium	37	5	32
Switzerland	32	6	26
Israel	31	5	26
Finland	30	1	29
Australia	26	3	23
Denmark	26	10	16
Scotland	25	9	16
India	23	0	23
China	22	1	21
Sweden	22	1	21
Greece	19	2	17

and strategies is of great importance for the learning and use of technology. (Martinez, Jaime and Camacho 2014) In this context, the importance of the increase in collaboration with the leading countries of the biotechnology such as USA, UK and Germany will be of vital importance, as well as the collaboration between countries sharing similar objectives and strategies. From this point of view, it would be appropriate to increase international collaboration with countries such as Argentina (4 co-operations), Brazil (12 co-operations), Republic of South Africa (15 co-operations) and South Korea (14 co-operations).

Table 5. Relation of Turkey's biotechnology publications to other research areas

Journals	Total	1973-2003	2004-2016	Impact Factor	Ranking in the Journals Listed under the Category of Biotechnology and Applied Microbiology
AFRICAN JOURNAL OF BIOTECHNOLOGY	684	0	684	0,573 (2010)	137/160 (2010)
GENETIC COUNSELING	485	57	428	0,236 (2016)	156/158 (2016)
BIORESOURCE TECHNOLOGY	380	45	335	5,651 (2016)	14/158 (2016)
PROCESS BIOCHEMISTRY	290	129	161	2,497 (2016)	64/158 (2016)
BIOTECHNOLOGY BIOTECHNOLOGICAL EQUIPMENT	264	45	219	1,059 (2016)	131/158 (2016)
JOURNAL OF CHEMICAL TECHNOLOGY AND BIOTECHNOLOGY	221	95	126	3,135 (2016)	48/158 (2016)
BIOMED RESEARCH INTERNATIONAL	185	0	185	2,476 (2016)	65/158 (2016)
ENZYME AND MICROBIAL TECHNOLOGY	154	84	70	2,502 (2016)	63/158 (2016)
APPLIED BIOCHEMISTRY AND BIOTECHNOLOGY	141	31	110	1,751 (2016)	94/158 (2016)
BIOTECHNIC HISTOCHEMISTRY	128	8	120	1,041 (2016)	133/158 (2016)
ANNALS OF MICROBIOLOGY	109	9	100	1,122 (2016)	129/158 (2016)
PREPARATIVE BIOCHEMISTRY BIOTECHNOLOGY	106	18	88	1,361 (2016)	118/158 (2016)
ARTIFICIAL CELLS NANOMEDICINE AND BIOTECHNOLOGY	98	0	98	5,605 (2016)	15/158 (2016)
BIOCHEMICAL ENGINEERING JOURNAL	97	25	72	2,892 (2016)	53/158 (2016)
CYTOTECHNOLOGY	97	2	95	1,857 (2016)	88/158 (2016)
ROMANIAN BIOTECHNOLOGICAL LETTERS	97	0	97	0,396 (2016)	150/158 (2016)
WORLD JOURNAL OF MICROBIOLOGY BIOTECHNOLOGY	96	22	74	1,658 (2016)	102/158 (2016)
MUTATION RESEARCH GENETIC TOXICOLOGY AND ENVIRONMENTAL MUTAGENESIS	84	31	53	2,133 (2016)	78/158 (2016)

continued on following page

Table 5. Continued

Journals	Total	1973-2003	2004-2016	Impact Factor	Ranking in the Journals Listed under the Category of Biotechnology and Applied Microbiology
BIOSENSORS BIOELECTRONICS	78	8	70	7,78 (2016)	9/158 (2016)
JOURNAL OF FOOD SAFETY	76	8	68	0,938 (2016)	145/158 (2016)
JOURNAL OF FOOD PROTECTION	74	17	57	1,417 (2016)	115/158 (2016)
ARTIFICIAL CELLS BLOOD SUBSTITUTES AND BIOTECHNOLOGY	70	4	66	1,314 (2014)	120/163 (2014)
BIOPROCESS AND BIOSYSTEMS ENGINEERING	64	0	64	1,87 (2016)	87/158 (2016)
INTERNATIONAL BIODETERIORATION BIODEGRADATION	62	4	58	2,962 (2016)	51/158 (2016)
JOURNAL OF BIOACTIVE AND COMPATIBLE POLYMERS	57	23	34	1,31 (2016)	122/158 (2016)

Journals in Which Studies Are Published and Subject Area

A total of 6,133 studies carried out by Turkey in the field of biotechnology in the period 1973-2016 were published in 178 journals. Table 5 contains the top 25 journals with the highest number of publications. 4,197 (68%) of 6,133 studies were published in these journals. The average impact factor of the top 25 journals published is 2,221, and the average impact factor of the 158 journals in the field of Biotechnology and Applied Microbiology as of 2016 is 3,282. This shows that the publications originating in Turkey concentrate on journals with a relatively lower impact factor.

The research areas of ISI Web of Science have been examined in order to reveal Turkey's biotechnology publications in relation to other disciplines. ISI research areas categorize the journals listed in the Science Citation Index. According to this, it is presented in Table 6 in which subject areas besides the subject field "Biotechnology & Applied Microbiology" of these 6133 publications Turkey has between 1973-2016. According to the table, Turkey's biotechnology publications concentrate the most in the engineering field.

Another remarkable point in Table 6 is the subject areas of publications in the 2004-2016 sub-period. According to this, Turkey's biotechnology

Table 6. The relation of Turkey's publications in the field of biotechnology with other subject areas

SUBJECT AREAS	1973-2003	2004-2016	Total
BIOTECHNOLOGY APPLIED MICROBIOLOGY	1086	5047	6133
ENGINEERING	304	690	994
BIOCHEMISTRY MOLECULAR BIOLOGY	244	604	848
RESEARCH EXPERIMENTAL MEDICINE	75	724	799
GENETICS HEREDITY	126	667	793
MEDICAL ETHICS	57	428	485
AGRICULTURE	73	409	482
MICROBIOLOGY	88	360	448
ENERGY FUELS	55	391	446
FOOD SCIENCE TECHNOLOGY	62	352	414
CHEMISTRY	109	208	317
CELL BIOLOGY	19	276	295
MATERIALS SCIENCE	44	219	263
TOXICOLOGY	48	79	127
SCIENCE TECHNOLOGY OTHER TOPICS	8	98	106
PLANT SCIENCES	23	58	81
BIOPHYSICS	8	70	78
ELECTROCHEMISTRY	8	70	78
IMMUNOLOGY	17	60	77
MATHEMATICAL COMPUTATIONAL BIOLOGY	2	69	71
ENVIRONMENTAL SCIENCES ECOLOGY	4	59	63
ONCOLOGY	18	43	61
POLYMER SCIENCE	23	35	58
ENTOMOLOGY	2	51	53
COMPUTER SCIENCE	2	43	45
MATHEMATICS	2	43	45
PATHOLOGY	2	42	44
VIROLOGY	6	32	38
HEMATOLOGY	3	30	33
MYCOLOGY	2	19	21
MARINE FRESHWATER BIOLOGY	2	14	16
PHARMACOLOGY PHARMACY	5	11	16
GENERAL INTERNAL MEDICINE	1	0	1
MEDICAL LABORATORY TECHNOLOGY	1	0	1
VETERINARY SCIENCES	1	0	1
ZOOLOGY	1	0	1

publications in the sub-period of 2004-2016 increased by 4.5 times compared to 1973-2003 sub-period whereas Mathematical Computational Biology increased by 34.5 times, Cell Biology by 14.5 times, Science Technology Other Topics by 12, 5 times and Research Experimental Medicine increased by 10 times. On the other hand, the rate of increase in the subjects such as Agriculture, Chemistry, Energy Fuels, Food Science Technology and Agriculture has been slower. This situation can be interpreted as a sign that the researches in the field of biotechnology in Turkey cannot be done sufficiently towards application areas. From this point of view, it has yet to be determined which traditional sectors biotechnology will focus on. It is necessary to establish and clearly define a strategy which sectors will be positioned on biotechnology and relevant policies need to be developed.

CONCLUSION

Turkey has an important potential in the field of biotechnology both in the production and commercialization of scientific knowledge. From this point of view, Turkey has started to recognize the importance of biotechnology from the middle of the 1980s, but a comprehensive strategy plan for this area was introduced only at the beginning of 2000s. The main aim of this study is to evaluate the information stocks and trends of Turkey in the field of biotechnology by using bibliometric research methods between 1973-2016. The study period was divided into two sub-periods (1973-2003 and 2004-2016) and Turkey's biotechnology trends, collaborations and research structure were analyzed. Thus, it is thought that the effects of the strategy plan that Turkey has introduced in 2000s can be seen more clearly. The lack of such a study in the field of biotechnology for Turkey has been the main motivation for this study.

When the general tendency of Turkey's biotechnology publications was evaluated, both the number of publications and the contribution to scientific knowledge accumulation of these publications increased significantly. However, the abovementioned increase is behind other emerging countries, which show a potential for growth in the field of biotechnology.

National/international collaborations are of great importance in terms of all new technologies in biotechnology. From this point of view, when two sub-periods are examined, the number of collaborative studies has increased both in absolute terms and proportionally. However, a significant part of these collaborations have been made between national institutions. Although the development of national collaborations is of great importance in the sense of information dissemination, the fact that international co-operations have not increased proportionally has led to a significant shortage of participation in international information networks.

When the first twenty national institutions that collaboration with one another are examined, it is seen that 19 are public universities and 1 is a public research institution. In terms of national collaborations, TÜBİTAK, a public research institution, has undertaken an important task. The fact that all but one of the first twenty institutions are public universities presents a significant shortage of actor diversity in terms of knowledge dissemination in the field of biotechnology. The national distribution of international collaboration seems to be concentrated in the top twenty countries. At this point, it is necessary to develop Turkey's collaboration with countries that share similar targets and strategies.

When it is examined in which journals and subject areas are concentrated the publications of Turkish origin, it is seen that the publication quality does not increase as expected and the expected concentration in the subject areas focusing on application cannot be achieved.

This study aimed to reveal the publication profile of Turkey in the field of biotechnology in the framework of records obtained from the ISI Web of Science database. As with every study, this study has important shortcomings and constraints. Above all, in order to obtain a more comprehensive publication profile, the publications covered by other national and international databases in addition to the articles cited in the database used in this study may be included in the analyzes. A similar research can also be done to commercialize biotechnology, including patent records. Finally, this study mainly aimed at defining the situation. The question of which policies can be passed on to address the deficiencies identified and the proposals made within the framework of these deficiencies requires a more comprehensive, historical and institutional research and therefore exceeds the scope of this study.

Despite all the shortcomings and limitations mentioned above, the findings of this study will guide other researchers, politicians, and more importantly, institutions that define visions and strategies in the field of biotechnology.

REFERENCES

Arencibia-Jorge, R., Corera-Alvarez, E., Chinchilla-Rodríguez, Z., & de Moya-Anegón, F. (2016). Scientific Output of Emerging Cuban Biopharmaceutical Industry: A Scientometric Approach. *Scientometrics*, *108*(3), 1621–1636. doi:10.100711192-016-2023-1

Bajwa, R. S., & Yaldram, K. (2013). Bibliometric Analysis of Biotechnology Research in Pakistan. *Scientometrics*, *95*(2), 529–540. doi:10.100711192-012-0839-x

Bartholomew, S. (1997). National Systems of Biotechnology Innovation: Complex Interdependence in the Global System. *Journal of International Business Studies*, *28*(2), 241–266. doi:10.1057/palgrave.jibs.8490100

Başağa, H. & Çetindamar, D. (2006). *Uluslararası Rekabet Stratejileri: Türkiye'de Biyoteknoloji Işbirlikleri*. Türkiye: Lebib Yalkın Yayımları ve Basım Işleri Anonim Şirketi.

Chen, K., & Guam, J. (2011). A bibliometric Investigation of Research Performance in Emerging Nanopharmaceuticals. *Journal of Informetrics*, *5*(2), 233–247. doi:10.1016/j.joi.2010.10.007

Costa, B. M. G., da Silva Pedro, E., & de Macedo, G. R. (2013). Scientific Collaboration in Biotechnology: The Case of The Northeast Region in Brazil. *Scientometrics*, *95*(2), 571–592. doi:10.100711192-012-0924-1

Erbaş, H. (2008). *Türkiye'de Biyoteknoloji ve Toplumsal Kesimler: Profesyoneller, Kentsel Tüketiciler ve Köylüler*. Ankara: Ankara University Basımevi.

Evans, P. (2010). Constructing the 21st century developmental state: Potentialities and pitfalls. Constructing the 21st century developmental state: Potentialities and pitfalls. In O. Edigheji (Ed.), *Constructing a democratic developmental state in south africa potentials and challenges* (pp. 37–58). HSRC Press.

Evans, P. (2012). What Will The Twenty First Century Developmental State Look Like? Implications of Contemporary Development Theory For The State's Role. In S. Wing-kai & W. Siu-lun (Eds.), *Repositioning the Hong Kong Government* (pp. 23–44). Hong Kong: Hong Kong University Press. doi:10.5790/hongkong/9789888083497.003.0002

Frenken, K., Hölzl, W., & de Vor, F. (2005). The Citation Impact of Research Collaborations: The Case of European Biotechnology and Applied Microbiology. *Journal of Engineering and Technology Management*, *22*(1-2), 9–30. doi:10.1016/j.jengtecman.2004.11.002

Glanzel, W., & Zhou, P. (2011). Publication Activity, Citation Impact and Bi-directional Links Between Publications and Patents in Biotechnology. *Scientometrics*, *86*(2), 505–525. doi:10.100711192-010-0269-6

Martinez, H., Jaime, A., & Camacho, J. (2014). Biotechnology Profile Analysis in Colombia. *Scientometrics*, *101*(3), 1789–1804. doi:10.100711192-014-1408-2

Molatudi, M., & Pouris, A. (2006). Assessing The Knowledge Base for Biotechnology in South Africa. *Scientometrics*, *68*(1), 97–108. doi:10.100711192-006-0085-1

Niosi, J. (2010). Rethinking Science, Technology and Innovation (STI) Institutions in Developing Countries. *Innovation: Management*, *Policy & Practice*, *12*(3), 250–268. doi:10.5172/impp.12.3.250

Niosi, J., Bas, T. G., & Amador, J. F. (2013). Biopharmaceuticals in Latin America: Challenges and Opportunities. *Innovation and Development*, *3*(1), 19–36. doi:10.1080/2157930X.2012.760899

Niosi, J., Hanel, P., & Reid, S. E. (2012). The International Diffusion of Biotechnology: The Arrival of Developing Countries. *Journal of Evolutionary Economics*, *22*(4), 767–783. doi:10.100700191-012-0284-2

Parakash, M., & Arumugam, J. (2017). Indian Contribution to Biotechnology Research: Scientometric Analysis. *Journal of Advances in Library and Information Science.*, *6*(1), 36–40.

Patra, S. (2005). Biotechnology Research Profile of India. *Scientometrics*, *63*(3), 583–597. doi:10.100711192-005-0229-8

Prange, H., & Kaiser, R. (2004). The Reconfiguration of National Innovation Systems- The Example of German Biotechnology. *Research Policy*, *33*(3), 395–408. doi:10.1016/j.respol.2003.09.001

Rasmussen, B. (2010). *Innovation and Commercialisation in the Biopharmaceutical Industry Creating and Capturing Value*. Cheltenham, UK: Edgar Elgar Publishing Limited. doi:10.4337/9781849805513

Rawat, H. (2008). *Textbook of Biotechnology*. Jaipur: Global Media.

T.C Bilim Teknoloji ve Sanayi Bakanlığı. (2015). *Türkiye Biyoteknoloji Stratejisi ve Eylem Planı 2015-2018*. Retrieved from http://www.sanayi.gov.tr/DokumanGetHandler.ashx?dokumanId=017882b9-01fe-4b8c-86dd-b5d9ca996e60

Tepav. (2015). *Yeni Sanayi Politikasının Odağında Teknoloji Platformları Var: Biyoteknoloji, Nanoteknoloji ve BÝT*. Türkiye Ekonomi Politikaları Araştırma Vakfı, Politika Notu: N201505.

Todeschini, R., & Baccini, A. (2016). *Handbook of Bibliometric Indicators*. Wiley.

Tunçgenç, M. (2014). *Dünyada ve Türkiye'de Biyoteknoloji Sanayisine Bakış*. Türkiye: Izmir.

Türkiye Bilimsel ve Teknik Araştırma Kurumu (Tübitak). (2004). *Ulusal Bilim ve Teknoloji Politikaları 2003-2023 Strateji Belgesi*. Retrieved from https://www.TÜBİTAK.gov.tr/TÜBİTAK_content_files//vizyon2023/Vizyon2023_Strateji_Belgesi.pdf

Wong, J. H., Chen, T., & Tsai, C. (2012). In Search of an Innovative State: The Development of the State of the Biopharmaceutical Industry in Taiwan, South Korea and China. *Development and Change*, *43*(2), 481–503. doi:10.1111/j.1467-7660.2012.01769.x

Yaşgül, Y. S., & Güriş, B. (2015). Causality between research output in the field of biotechnology and economic growth in Turkey. *Quality & Quantity*, *50*(4), 1715–1726. doi:10.100711135-015-0230-0

Zhao, Y., Li, D., Han, M., Li, C., & Li, D. (2016). Characteristics of Research Collaboration in Biotechnology in China: Evidence from Publications. *Scientometrics*, *107*(3), 1373–1387. doi:10.100711192-016-1898-1

ENDNOTES

1 The activity index is calculated as the ratio of N countries in terms of all publications in the world in the F field of research to the share of N

countries in terms of all publications in the world in all research fields. Accordingly, when the activity index is greater than 1, F refers to a relative specialty in the field of research. For detailed information, see Todeschini and Baccini (2016).

2 The average citation per publication is obtained by dividing the total number of citations made to all publications by the total number of publications. It is assumed that as the average number of citations increases, the contribution of publications to scientific knowledge increases.

3 In any publication involving national collaboration, the number of collaboration is more than the number of publications, as there can be more than one national institution.

4 Due to limited space, Table 2 is limited to the first twenty national institutions. Only 1 private sector research institute is located within the top 100 national institutions.

Related Readings

To continue IGI Global's long-standing tradition of advancing innovation through emerging research, please find below a compiled list of recommended IGI Global book chapters and journal articles in the areas of business leadership, entrepreneurship, and gulf cooperation council. These related readings will provide additional information and guidance to further enrich your knowledge and assist you with your own research.

Abalkhail, J. M. (2018). Arab Gulf Women and the Labyrinth of Leadership. In E. Al-A'ali, M. Al-Shammari, & H. Masri (Eds.), *Arab Women and Their Evolving Roles in the Global Business Landscape* (pp. 26–50). Hershey, PA: IGI Global. doi:10.4018/978-1-5225-3710-6.ch002

Abe, E. N., & Abe, I. I. (2019). Integrating Multiculturalism in the Design and Implementation of Work-Life Balance Strategies. In S. Atiku (Ed.), *Contemporary Multicultural Orientations and Practices for Global Leadership* (pp. 100–120). Hershey, PA: IGI Global. doi:10.4018/978-1-5225-6286-3.ch006

Abe, I. I., & Abe, E. N. (2019). Dynamics of Human Resource Strategies and Cultural Orientations in Multinational Corporations. In S. Atiku (Ed.), *Contemporary Multicultural Orientations and Practices for Global Leadership* (pp. 80–99). Hershey, PA: IGI Global. doi:10.4018/978-1-5225-6286-3.ch005

Ak, R., & Bingül, B. A. (2018). Kuwait. In S. Ozdemir, S. Erdogan, & A. Gedikli (Eds.), *Handbook of Research on Sociopolitical Factors Impacting Economic Growth in Islamic Nations* (pp. 276–292). Hershey, PA: IGI Global. doi:10.4018/978-1-5225-2939-2.ch014

Akkoyunlu, P. F. (2016). Assessing the Effectiveness of Vocational Education and Training in Economic Development: A Comparison between Turkish and German Education Systems. In M. Erdoğdu & B. Christiansen (Eds.), *Handbook of Research on Comparative Economic Development Perspectives on Europe and the MENA Region* (pp. 127–151). Hershey, PA: IGI Global. doi:10.4018/978-1-4666-9548-1.ch007

Akram, M. U., Chauhan, C., Ghosh, K., & Singh, A. (2019). Knowledge Management, Sustainable Business Performance and Empowering Leadership: A Firm-Level Approach. *International Journal of Knowledge Management*, *15*(2), 20–35. doi:10.4018/IJKM.2019040102

Al-Alawi, A. I., Husamaddin, S., Mejeran, F. K., & Madan, F. K. (2018). Bahraini Women Engineers: Factors Influencing Their Career Success Ladder. In E. Al-A'ali, M. Al-Shammari, & H. Masri (Eds.), *Arab Women and Their Evolving Roles in the Global Business Landscape* (pp. 178–208). Hershey, PA: IGI Global. doi:10.4018/978-1-5225-3710-6.ch008

Al-Yafi, K., Hindi, N. M., & Osman, I. H. (2016). A User-Centric Evaluation of e-Government Services in the GCC Region: Case of State of Qatar. *International Journal of Electronic Government Research*, *12*(4), 15–34. doi:10.4018/IJEGR.2016100102

Alfantookh, A., & Bakry, S. H. (2017). The State of People and Knowledge in the GCC Countries per a New Index and the Future Ahead. *International Journal of Knowledge Society Research*, *8*(4), 34–74. doi:10.4018/IJKSR.2017100103

Alkhuraiji, A. I. (2018). The Influence of the Top Management Characteristics on the Success of the Enterprise Information System. In A. Borchers (Ed.), *Technology Management in Organizational and Societal Contexts* (pp. 1–26). Hershey, PA: IGI Global. doi:10.4018/978-1-5225-5279-6.ch001

Alkraiji, A. I. (2016). Cross-Case Analysis of Top Management Characteristics and Enterprise Information System Success. *Journal of Cases on Information Technology*, *18*(3), 40–58. doi:10.4018/JCIT.2016070104

Alraouf, A. A. (2018). A Holistic Assessment of Education City in Doha: The Borderless Knowledge Hub. In A. Alraouf (Ed.), *Knowledge-Based Urban Development in the Middle East* (pp. 228–247). Hershey, PA: IGI Global. doi:10.4018/978-1-5225-3734-2.ch012

Asiri, M. (2017). The Interaction of Pharmaceutical Regulations in Saudi Arabia With the Presence of FDI: Facilitating FDI in Saudi Pharmaceutical Industry. In B. Christiansen & G. Koc (Eds.), *Transcontinental Strategies for Industrial Development and Economic Growth* (pp. 132–145). Hershey, PA: IGI Global. doi:10.4018/978-1-5225-2160-0.ch008

Atiku, S. O. (2019). Institutionalizing Social Responsibility Through Workplace Green Behavior. In S. Atiku (Ed.), *Contemporary Multicultural Orientations and Practices for Global Leadership* (pp. 183–199). Hershey, PA: IGI Global. doi:10.4018/978-1-5225-6286-3.ch010

Atiku, S. O., & Fields, Z. (2019). Global Psychological Capital and Sustainable Competitive Advantage. In S. Atiku (Ed.), *Contemporary Multicultural Orientations and Practices for Global Leadership* (pp. 145–164). Hershey, PA: IGI Global. doi:10.4018/978-1-5225-6286-3.ch008

Atsan, N. (2019). Utilizing Social Networks in Social Entrepreneurship. In N. Iyigun (Ed.), *Creating Business Value and Competitive Advantage With Social Entrepreneurship* (pp. 232–250). Hershey, PA: IGI Global. doi:10.4018/978-1-5225-5687-9.ch011

Ayandibu, A. O. (2019). Green Transformational Leadership and Organizational Behavior Effectiveness. In S. Atiku (Ed.), *Contemporary Multicultural Orientations and Practices for Global Leadership* (pp. 225–244). Hershey, PA: IGI Global. doi:10.4018/978-1-5225-6286-3.ch012

Bartens, Y., Chunpir, H. I., Schulte, F., & Voß, S. (2017). Business/IT Alignment in Two-Sided Markets: A COBIT 5 Analysis for Media Streaming Business Models. In S. De Haes & W. Van Grembergen (Eds.), *Strategic IT Governance and Alignment in Business Settings* (pp. 82–111). Hershey, PA: IGI Global. doi:10.4018/978-1-5225-0861-8.ch004

Bayarçelik, E. B., & Taşel, F. (2019). Social Entrepreneurship for Sustainable Economic Growth. In N. Iyigun (Ed.), *Creating Business Value and Competitive Advantage With Social Entrepreneurship* (pp. 251–270). Hershey, PA: IGI Global. doi:10.4018/978-1-5225-5687-9.ch012

Bayik, M. F. (2019). Social Innovation and Social Entrepreneurship. In N. Iyigun (Ed.), *Creating Business Value and Competitive Advantage With Social Entrepreneurship* (pp. 92–114). Hershey, PA: IGI Global. doi:10.4018/978-1-5225-5687-9.ch005

Ben Hamida, L. (2016). Building R&D Capabilities Abroad and the Role of Reverse Knowledge Transfer in Explaining MNCs' Productivity. In M. Erdoğdu & B. Christiansen (Eds.), *Handbook of Research on Comparative Economic Development Perspectives on Europe and the MENA Region* (pp. 219–235). Hershey, PA: IGI Global. doi:10.4018/978-1-4666-9548-1.ch011

Bergeron, F., Croteau, A., Uwizeyemungu, S., & Raymond, L. (2017). A Framework for Research on Information Technology Governance in SMEs. In S. De Haes & W. Van Grembergen (Eds.), *Strategic IT Governance and Alignment in Business Settings* (pp. 53–81). Hershey, PA: IGI Global. doi:10.4018/978-1-5225-0861-8.ch003

Bergethon, K. P., & Davis, D. C. (2016). Emotional Leadership: Leadership Styles and Emotional Intelligence. In A. Normore, L. Long, & M. Javidi (Eds.), *Handbook of Research on Effective Communication, Leadership, and Conflict Resolution* (pp. 63–77). Hershey, PA: IGI Global. doi:10.4018/978-1-4666-9970-0.ch004

Bhattacharya, P. J. (2018). Managing IT to Innovate and Strategize in the Higher Education Sector: Role of Enterprise Systems. In A. Borchers (Ed.), *Technology Management in Organizational and Societal Contexts* (pp. 27–50). Hershey, PA: IGI Global. doi:10.4018/978-1-5225-5279-6.ch002

Butt, T. A., Iqbal, R., & Kehal, M. (2019). Social Internet of Vehicles: A GCC (Gulf Cooperation Council) Perspective. In Z. Zhang (Ed.), *Global Information Diffusion and Management in Contemporary Society* (pp. 172–188). Hershey, PA: IGI Global. doi:10.4018/978-1-5225-5393-9.ch007

Camillo, A. A., & Camillo, I. C. (2016). The Ethics of Strategic Managerial Communication in the Global Context. In A. Normore, L. Long, & M. Javidi (Eds.), *Handbook of Research on Effective Communication, Leadership, and Conflict Resolution* (pp. 566–590). Hershey, PA: IGI Global. doi:10.4018/978-1-4666-9970-0.ch030

Carlos, V. S., & Rodrigues, R. G. (2016). The Use of Online Social Networks in Higher Education and Its Influence on Job Performance. In A. Normore, L. Long, & M. Javidi (Eds.), *Handbook of Research on Effective Communication, Leadership, and Conflict Resolution* (pp. 330–353). Hershey, PA: IGI Global. doi:10.4018/978-1-4666-9970-0.ch018

Cavusoglu, T., & Onal, D. K. (2016). The Role of Institutional Quality on Total Factor Productivity: A Comparative Analysis on the CEE and MENA Economies. In M. Erdoğdu & B. Christiansen (Eds.), *Handbook of Research on Comparative Economic Development Perspectives on Europe and the MENA Region* (pp. 98–126). Hershey, PA: IGI Global. doi:10.4018/978-1-4666-9548-1.ch006

Chen, E. T. (2016). Virtual Team Management for Higher Performance. In A. Normore, L. Long, & M. Javidi (Eds.), *Handbook of Research on Effective Communication, Leadership, and Conflict Resolution* (pp. 298–310). Hershey, PA: IGI Global. doi:10.4018/978-1-4666-9970-0.ch016

Clark, C., & Stowers, G. (2016). Speaking with Trunks, Dancing with the "Pink Elephants": Troubling E-Racism, E-Classism, and E-Sexismin Teaching Multicultural Teacher Education. In A. Normore, L. Long, & M. Javidi (Eds.), *Handbook of Research on Effective Communication, Leadership, and Conflict Resolution* (pp. 78–97). Hershey, PA: IGI Global. doi:10.4018/978-1-4666-9970-0.ch005

Clarke, M., Hall, J. G., & Rapanotti, L. (2017). Enterprise Architecture: A Snapshot from Practice. In S. De Haes & W. Van Grembergen (Eds.), *Strategic IT Governance and Alignment in Business Settings* (pp. 237–255). Hershey, PA: IGI Global. doi:10.4018/978-1-5225-0861-8.ch009

Cockburn, T., & Smith, P. A. (2016). Leadership in the Digital Age: Rhythms and the Beat of Change. In A. Normore, L. Long, & M. Javidi (Eds.), *Handbook of Research on Effective Communication, Leadership, and Conflict Resolution* (pp. 1–20). Hershey, PA: IGI Global. doi:10.4018/978-1-4666-9970-0.ch001

Cockburn, T., Smith, P. A., Martins, B. M., & Valles, R. S. (2016). Conflict of Interest or Community of Collaboration?: Leadership, SME Network Dialectics, and Dialogs. In A. Normore, L. Long, & M. Javidi (Eds.), *Handbook of Research on Effective Communication, Leadership, and Conflict Resolution* (pp. 210–232). Hershey, PA: IGI Global. doi:10.4018/978-1-4666-9970-0.ch012

Connolly, R., & Kenny, G. (2016). Dataveillance and Information Privacy Concerns: Ethical and Organizational Considerations. In A. Normore, L. Long, & M. Javidi (Eds.), *Handbook of Research on Effective Communication, Leadership, and Conflict Resolution* (pp. 606–626). Hershey, PA: IGI Global. doi:10.4018/978-1-4666-9970-0.ch032

Cross, D. E. (2016). Globalization and Media's Impact on Cross Cultural Communication: Managing Organizational Change. In A. Normore, L. Long, & M. Javidi (Eds.), *Handbook of Research on Effective Communication, Leadership, and Conflict Resolution* (pp. 21–41). Hershey, PA: IGI Global. doi:10.4018/978-1-4666-9970-0.ch002

Dahlberg, T., Kivijärvi, H., & Saarinen, T. (2017). IT Investment Consistency and Other Factors Influencing the Success of IT Performance. In S. De Haes & W. Van Grembergen (Eds.), *Strategic IT Governance and Alignment in Business Settings* (pp. 176–208). Hershey, PA: IGI Global. doi:10.4018/978-1-5225-0861-8.ch007

Davis, D. C., & Scaffidi-Clarke, N. M. (2016). Leading Virtual Teams: Conflict and Communication Challenges for Leaders. In A. Normore, L. Long, & M. Javidi (Eds.), *Handbook of Research on Effective Communication, Leadership, and Conflict Resolution* (pp. 196–209). Hershey, PA: IGI Global. doi:10.4018/978-1-4666-9970-0.ch011

Duarte, M., Baptista, A., & Pinto, G. (2018). Learning in the Laboratory: Accessing Videos With Quick Response Codes. In A. Borchers (Ed.), *Technology Management in Organizational and Societal Contexts* (pp. 117–138). Hershey, PA: IGI Global. doi:10.4018/978-1-5225-5279-6.ch006

Durgun, Ö. (2018). United Arab Emirates. In S. Ozdemir, S. Erdogan, & A. Gedikli (Eds.), *Handbook of Research on Sociopolitical Factors Impacting Economic Growth in Islamic Nations* (pp. 253–275). Hershey, PA: IGI Global. doi:10.4018/978-1-5225-2939-2.ch013

El-Kholei, A. O., & Al-Jayyousi, O. (2018). Cities of GCC in Post-Hydrocarbon Era: Challenges and Opportunities. In A. Alraouf (Ed.), *Knowledge-Based Urban Development in the Middle East* (pp. 248–270). Hershey, PA: IGI Global. doi:10.4018/978-1-5225-3734-2.ch013

Ellis, B., & Normore, A. H. (2016). Effective Engagement and Communication between First-Line Police Supervisors and Police Officers. In A. Normore, L. Long, & M. Javidi (Eds.), *Handbook of Research on Effective Communication, Leadership, and Conflict Resolution* (pp. 479–493). Hershey, PA: IGI Global. doi:10.4018/978-1-4666-9970-0.ch025

Erdoğan, S., & Gedikli, A. (2018). The State of Qatar. In S. Ozdemir, S. Erdogan, & A. Gedikli (Eds.), *Handbook of Research on Sociopolitical Factors Impacting Economic Growth in Islamic Nations* (pp. 212–252). Hershey, PA: IGI Global. doi:10.4018/978-1-5225-2939-2.ch012

Ertemel, A. V. (2019). Lean Approach to Social Entrepreneurship. In N. Iyigun (Ed.), *Creating Business Value and Competitive Advantage With Social Entrepreneurship* (pp. 115–139). Hershey, PA: IGI Global. doi:10.4018/978-1-5225-5687-9.ch006

Ezziane, Z., & Al Shamisi, A. (2017). The Impact of IT Governance Compliance on Enhancing Organizational Performance in Abu Dhabi. In S. De Haes & W. Van Grembergen (Eds.), *Strategic IT Governance and Alignment in Business Settings* (pp. 209–236). Hershey, PA: IGI Global. doi:10.4018/978-1-5225-0861-8.ch008

Farrag, N. A., & Ezzat, A. M. (2016). The Impact of Corruption on Economic Growth: A Comparative Analysis between Europe and MENA Countries. In M. Erdoğdu & B. Christiansen (Eds.), *Handbook of Research on Comparative Economic Development Perspectives on Europe and the MENA Region* (pp. 74–97). Hershey, PA: IGI Global. doi:10.4018/978-1-4666-9548-1.ch005

Fernandes, V. (2016). Reframing Continuous School Improvement in Australian Schools. In A. Normore, L. Long, & M. Javidi (Eds.), *Handbook of Research on Effective Communication, Leadership, and Conflict Resolution* (pp. 98–124). Hershey, PA: IGI Global. doi:10.4018/978-1-4666-9970-0.ch006

Fernández, M. T., & Pablo-Marti, F. (2016). Industrial Competitiveness in MENA Countries: Current Strategic Directions of Industrial Policy. In M. Erdoğdu & B. Christiansen (Eds.), *Handbook of Research on Comparative Economic Development Perspectives on Europe and the MENA Region* (pp. 152–171). Hershey, PA: IGI Global. doi:10.4018/978-1-4666-9548-1.ch008

Fındıklı, M. A., & Yozgat, U. (2019). The Individual and Behavioral Factors of Social Entrepreneurs. In N. Iyigun (Ed.), *Creating Business Value and Competitive Advantage With Social Entrepreneurship* (pp. 69–91). Hershey, PA: IGI Global. doi:10.4018/978-1-5225-5687-9.ch004

Genty, K. I. (2019). Interpersonal Relationship and Global Leadership Mindset. In S. Atiku (Ed.), *Contemporary Multicultural Orientations and Practices for Global Leadership* (pp. 165–182). Hershey, PA: IGI Global. doi:10.4018/978-1-5225-6286-3.ch009

Gomes, J., & Romão, M. (2017). Aligning Information Systems and Technology with Benefit Management and Balanced Scorecard. In S. De Haes & W. Van Grembergen (Eds.), *Strategic IT Governance and Alignment in Business Settings* (pp. 112–131). Hershey, PA: IGI Global. doi:10.4018/978-1-5225-0861-8.ch005

Gul, S. (2016). Alternative Dispute Resolution: A Legal Perspective. In A. Normore, L. Long, & M. Javidi (Eds.), *Handbook of Research on Effective Communication, Leadership, and Conflict Resolution* (pp. 126–143). Hershey, PA: IGI Global. doi:10.4018/978-1-4666-9970-0.ch007

Güler, B. K. (2019). Human Resources Management in Social Entrepreneurship. In N. Iyigun (Ed.), *Creating Business Value and Competitive Advantage With Social Entrepreneurship* (pp. 161–185). Hershey, PA: IGI Global. doi:10.4018/978-1-5225-5687-9.ch008

Gupta, D. S., & Biswas, G. P. (2018). On Securing Cloud Storage Using a Homomorphic Framework. In A. Borchers (Ed.), *Technology Management in Organizational and Societal Contexts* (pp. 99–114). Hershey, PA: IGI Global. doi:10.4018/978-1-5225-5279-6.ch005

Huda, M., Qodriah, S. L., Rismayadi, B., Hananto, A., Kardiyati, E. N., Ruskam, A., & Nasir, B. M. (2019). Towards Cooperative With Competitive Alliance: Insights Into Performance Value in Social Entrepreneurship. In N. Iyigun (Ed.), *Creating Business Value and Competitive Advantage With Social Entrepreneurship* (pp. 294–317). Hershey, PA: IGI Global. doi:10.4018/978-1-5225-5687-9.ch014

Igbinakhase, I., & Naidoo, V. (2019). Global Leadership Competencies and Orientations: Theory, Practice, and Future Implications. In S. Atiku (Ed.), *Contemporary Multicultural Orientations and Practices for Global Leadership* (pp. 59–79). Hershey, PA: IGI Global. doi:10.4018/978-1-5225-6286-3.ch004

Inglezakis, I. (2016). Surveillance of Electronic Communications in the Workplace and the Protection of Employees' Privacy. In A. Normore, L. Long, & M. Javidi (Eds.), *Handbook of Research on Effective Communication, Leadership, and Conflict Resolution* (pp. 591–605). Hershey, PA: IGI Global. doi:10.4018/978-1-4666-9970-0.ch031

İyigün, N. Ö. (2019). Context of Social Entrepreneurship and Background: Definitions, Related Terms, and Characteristics. In N. Iyigun (Ed.), *Creating Business Value and Competitive Advantage With Social Entrepreneurship* (pp. 1–20). Hershey, PA: IGI Global. doi:10.4018/978-1-5225-5687-9.ch001

Jackson, K., & Rasheed, R. (2016). Communication, Culture, and Discord: A Lesson in Leadership Failure. In A. Normore, L. Long, & M. Javidi (Eds.), *Handbook of Research on Effective Communication, Leadership, and Conflict Resolution* (pp. 182–195). Hershey, PA: IGI Global. doi:10.4018/978-1-4666-9970-0.ch010

Kabadurmus, F. N., & Lahiri, S. (2016). Competition and Product Innovation by Turkish Firms. In M. Erdoğdu & B. Christiansen (Eds.), *Handbook of Research on Comparative Economic Development Perspectives on Europe and the MENA Region* (pp. 469–491). Hershey, PA: IGI Global. doi:10.4018/978-1-4666-9548-1.ch019

Kaidalova, J., & Seigerroth, U. (2017). Contemporary Demands on Business and IT Alignment: How Can Enterprise Modeling Help? In S. De Haes & W. Van Grembergen (Eds.), *Strategic IT Governance and Alignment in Business Settings* (pp. 132–175). Hershey, PA: IGI Global. doi:10.4018/978-1-5225-0861-8.ch006

Kalitanyi, V. (2019). Global Business Environment. In S. Atiku (Ed.), *Contemporary Multicultural Orientations and Practices for Global Leadership* (pp. 1–24). Hershey, PA: IGI Global. doi:10.4018/978-1-5225-6286-3.ch001

Karaca, M., & Biçkes, D. M. (2019). Social Entrepreneurship: Constraints and Suggestions for Future Researches. In N. Iyigun (Ed.), *Creating Business Value and Competitive Advantage With Social Entrepreneurship* (pp. 318–332). Hershey, PA: IGI Global. doi:10.4018/978-1-5225-5687-9.ch015

Kariuki, P. (2019). Monitoring and Evaluation Leadership Through Technology: The South African Public-Sector Perspective. In S. Atiku (Ed.), *Contemporary Multicultural Orientations and Practices for Global Leadership* (pp. 121–144). Hershey, PA: IGI Global. doi:10.4018/978-1-5225-6286-3.ch007

Karnaukhova, O., & Nekrasova, I. (2016). The Role of Securitization Market in the Post-Crisis European Economic Recovery. In M. Erdoğdu & B. Christiansen (Eds.), *Handbook of Research on Comparative Economic Development Perspectives on Europe and the MENA Region* (pp. 511–523). Hershey, PA: IGI Global. doi:10.4018/978-1-4666-9548-1.ch021

Kayser, I. (2018). Who Benefits From Regional Economic Integration?: The Case of ASEAN. In A. Özer (Ed.), *Globalization and Trade Integration in Developing Countries* (pp. 97–115). Hershey, PA: IGI Global. doi:10.4018/978-1-5225-4032-8.ch005

Keil, L. J., & Jerome, A. M. (2016). Leadership in a Time of Crisis: Jim Tressel's Ousting from The Ohio State University. In A. Normore, L. Long, & M. Javidi (Eds.), *Handbook of Research on Effective Communication, Leadership, and Conflict Resolution* (pp. 162–181). Hershey, PA: IGI Global. doi:10.4018/978-1-4666-9970-0.ch009

Khoury, G. C., & McNally, B. (2016). The Role of the Leadership Style in Creating Conflict and Tension in a Higher Education Institution. In A. Normore, L. Long, & M. Javidi (Eds.), *Handbook of Research on Effective Communication, Leadership, and Conflict Resolution* (pp. 233–247). Hershey, PA: IGI Global. doi:10.4018/978-1-4666-9970-0.ch013

Kirstein, K. D. (2016). Developing Trust within International Teams: The Impacts of Culture on Team Formation and Process. In A. Normore, L. Long, & M. Javidi (Eds.), *Handbook of Research on Effective Communication, Leadership, and Conflict Resolution* (pp. 385–404). Hershey, PA: IGI Global. doi:10.4018/978-1-4666-9970-0.ch021

Long, L. W., Javidi, M., Hill, L. B., & Normore, A. H. (2016). Credible Negotiation Leadership: Using Principled Negotiation to Improve International Negotiation. In A. Normore, L. Long, & M. Javidi (Eds.), *Handbook of Research on Effective Communication, Leadership, and Conflict Resolution* (pp. 430–455). Hershey, PA: IGI Global. doi:10.4018/978-1-4666-9970-0.ch023

Maes, K., De Haes, S., & Van Grembergen, W. (2017). A Business Case Process for IT-Enabled Investments: Its Perceived Effectiveness from a Practitioner Perspective. In S. De Haes & W. Van Grembergen (Eds.), *Strategic IT Governance and Alignment in Business Settings* (pp. 1–23). Hershey, PA: IGI Global. doi:10.4018/978-1-5225-0861-8.ch001

Maher, C. (2019). Public Services Procurement: Third Sector Social Enterprises Perspectives. In N. Iyigun (Ed.), *Creating Business Value and Competitive Advantage With Social Entrepreneurship* (pp. 352–373). Hershey, PA: IGI Global. doi:10.4018/978-1-5225-5687-9.ch017

Manzoor, A. (2017). Accelerating Entrepreneurship in MENA Region: Opportunities and Challenges. In P. Zgheib (Ed.), *Entrepreneurship and Business Innovation in the Middle East* (pp. 57–73). Hershey, PA: IGI Global. doi:10.4018/978-1-5225-2066-5.ch004

Manzoor, A. (2017). Womenpreneurs in Mena Region. *International Journal of E-Entrepreneurship and Innovation*, *7*(1), 16–28. doi:10.4018/IJEEI.2017010102

Manzoor, A. (2018). High-Growth Entrepreneurship in the MENA Region. *International Journal of ICT Research in Africa and the Middle East*, *7*(1), 54–65. doi:10.4018/IJICTRAME.2018010104

Marois, T. (2016). State-Owned Banks and Development: Dispelling Mainstream Myths. In M. Erdoğdu & B. Christiansen (Eds.), *Handbook of Research on Comparative Economic Development Perspectives on Europe and the MENA Region* (pp. 52–73). Hershey, PA: IGI Global. doi:10.4018/978-1-4666-9548-1.ch004

Mishra, K. E., Mishra, A. K., & Walker, K. (2016). Leadership Communication, Internal Marketing, and Employee Engagement: A Recipe to Create Brand Ambassadors. In A. Normore, L. Long, & M. Javidi (Eds.), *Handbook of Research on Effective Communication, Leadership, and Conflict Resolution* (pp. 311–329). Hershey, PA: IGI Global. doi:10.4018/978-1-4666-9970-0.ch017

Mitchell, R. J., & Von Zoller, K. (2016). The Link between Communicative Intelligence and Procedural Justice: The Path to Police Legitimacy. In A. Normore, L. Long, & M. Javidi (Eds.), *Handbook of Research on Effective Communication, Leadership, and Conflict Resolution* (pp. 456–478). Hershey, PA: IGI Global. doi:10.4018/978-1-4666-9970-0.ch024

Moner-Colonques, R., & Sempere-Monerris, J. J. (2016). R&D Competition, Cooperation, and Microeconomic Policies. In M. Erdoğdu & B. Christiansen (Eds.), *Handbook of Research on Comparative Economic Development Perspectives on Europe and the MENA Region* (pp. 195–218). Hershey, PA: IGI Global. doi:10.4018/978-1-4666-9548-1.ch010

Mukonza, C., & Swarts, I. (2019). Examining the Role of Green Transformational Leadership on Promoting Green Organizational Behavior. In S. Atiku (Ed.), *Contemporary Multicultural Orientations and Practices for Global Leadership* (pp. 200–224). Hershey, PA: IGI Global. doi:10.4018/978-1-5225-6286-3.ch011

Nour, S. S. (2016). Characteristics of Innovation in the Arabic Countries. In L. Al-Hakim, X. Wu, A. Koronios, & Y. Shou (Eds.), *Handbook of Research on Driving Competitive Advantage through Sustainable, Lean, and Disruptive Innovation* (pp. 614–655). Hershey, PA: IGI Global. doi:10.4018/978-1-5225-0135-0.ch025

Nylund, P. A., Arimany-Serrat, N., Ferràs-Hernández, X., & Corral-Marfil, J. A. (2018). Home Bias in Innovation Ecosystems: Relying on Local Users for Knowledge. In A. Borchers (Ed.), *Technology Management in Organizational and Societal Contexts* (pp. 78–97). Hershey, PA: IGI Global. doi:10.4018/978-1-5225-5279-6.ch004

Öncer, A. Z. (2019). Performance Measurement in Social Enterprises: Social Impact Analysis. In N. Iyigun (Ed.), *Creating Business Value and Competitive Advantage With Social Entrepreneurship* (pp. 205–231). Hershey, PA: IGI Global. doi:10.4018/978-1-5225-5687-9.ch010

Önder, I., & Uckac, A. (2016). Opportunities and Challenges for Development. In M. Erdoğdu & B. Christiansen (Eds.), *Handbook of Research on Comparative Economic Development Perspectives on Europe and the MENA Region* (pp. 36–51). Hershey, PA: IGI Global. doi:10.4018/978-1-4666-9548-1.ch003

Oniku, A., Emmanuel, A. A., & Adetunji, O. (2017). Ethical Leadership, Ethical Selling, and Financial Services Performances: A Case of Nigerian Market. *International Journal of Sustainable Economies Management, 6*(4), 21–34. doi:10.4018/IJSEM.2017100103

Oz, S. (2016). Diffusion of Technology via FDI and Convergence of Per Capita Incomes: Comparative Analysis on Europe and the MENA Region. In M. Erdoğdu & B. Christiansen (Eds.), *Handbook of Research on Comparative Economic Development Perspectives on Europe and the MENA Region* (pp. 236–264). Hershey, PA: IGI Global. doi:10.4018/978-1-4666-9548-1.ch012

Ozer, M., & Yeldan, A. E. (2016). The Relationship between Current Account Deficits and Unemployment in Turkey. In M. Erdoğdu & B. Christiansen (Eds.), *Handbook of Research on Comparative Economic Development Perspectives on Europe and the MENA Region* (pp. 492–510). Hershey, PA: IGI Global. doi:10.4018/978-1-4666-9548-1.ch020

Ozer, S. K., Taglioni, D., & Winkler, D. (2016). Turkey's Participation and Economic Upgrading in Global Value Chains. In M. Erdoğdu & B. Christiansen (Eds.), *Handbook of Research on Comparative Economic Development Perspectives on Europe and the MENA Region* (pp. 418–468). Hershey, PA: IGI Global. doi:10.4018/978-1-4666-9548-1.ch018

Park, H. J. (2016). Recasting Strategic Opportunities and Challenges for Development in Saudi Arabia. In M. Erdoğdu & B. Christiansen (Eds.), *Handbook of Research on Comparative Economic Development Perspectives on Europe and the MENA Region* (pp. 406–417). Hershey, PA: IGI Global. doi:10.4018/978-1-4666-9548-1.ch017

Pereira, R., Mira da Silva, M., & Lapão, L. V. (2017). IT Governance Maturity Patterns in Portuguese Healthcare. In S. De Haes & W. Van Grembergen (Eds.), *Strategic IT Governance and Alignment in Business Settings* (pp. 24–52). Hershey, PA: IGI Global. doi:10.4018/978-1-5225-0861-8.ch002

Perez-Montoro, M., & Sanz, S. (2016). Communities of Practice in Organizational Learning Strategies. In A. Normore, L. Long, & M. Javidi (Eds.), *Handbook of Research on Effective Communication, Leadership, and Conflict Resolution* (pp. 249–266). Hershey, PA: IGI Global. doi:10.4018/978-1-4666-9970-0.ch014

Popovic, M. (2016). Causes of Growth of MENA and East European Countries: Comparative Analysis. In M. Erdoğdu & B. Christiansen (Eds.), *Handbook of Research on Comparative Economic Development Perspectives on Europe and the MENA Region* (pp. 293–350). Hershey, PA: IGI Global. doi:10.4018/978-1-4666-9548-1.ch014

Pyle, A. S. (2016). Surviving the Conflict of Self-Inflicted Organizational Crises. In A. Normore, L. Long, & M. Javidi (Eds.), *Handbook of Research on Effective Communication, Leadership, and Conflict Resolution* (pp. 144–161). Hershey, PA: IGI Global. doi:10.4018/978-1-4666-9970-0.ch008

Radzuan, N. R., & Kaur, S. (2016). Engineering Students' Communication Apprehension and Competence in Technical Oral Presentations. In A. Normore, L. Long, & M. Javidi (Eds.), *Handbook of Research on Effective Communication, Leadership, and Conflict Resolution* (pp. 371–383). Hershey, PA: IGI Global. doi:10.4018/978-1-4666-9970-0.ch020

Rafay, A., & Khan, A. (2018). Synergy for Sustainability in the Upcoming Telecommunications Revolution: The Case of a Developing Economy. In A. Borchers (Ed.), *Technology Management in Organizational and Societal Contexts* (pp. 51–76). Hershey, PA: IGI Global. doi:10.4018/978-1-5225-5279-6.ch003

Rahim, E., & Duncan, T. (2019). Examining the Evolution of Creating Sustainable Leadership. In S. Atiku (Ed.), *Contemporary Multicultural Orientations and Practices for Global Leadership* (pp. 44–58). Hershey, PA: IGI Global. doi:10.4018/978-1-5225-6286-3.ch003

Ramraj, A. B., & Amolo, J. (2019). Leadership Across the Globe. In S. Atiku (Ed.), *Contemporary Multicultural Orientations and Practices for Global Leadership* (pp. 25–43). Hershey, PA: IGI Global. doi:10.4018/978-1-5225-6286-3.ch002

Razzak, W. A., Laabas, B., & Bentour, E. M. (2016). The Dynamics of Technical Progress in Some Developing and Developed Countries. In M. Erdoğdu & B. Christiansen (Eds.), *Handbook of Research on Comparative Economic Development Perspectives on Europe and the MENA Region* (pp. 172–194). Hershey, PA: IGI Global. doi:10.4018/978-1-4666-9548-1.ch009

Rivlin, P. (2016). Stabilization and Liberalization in the Israeli Economy. In M. Erdoğdu & B. Christiansen (Eds.), *Handbook of Research on Comparative Economic Development Perspectives on Europe and the MENA Region* (pp. 524–540). Hershey, PA: IGI Global. doi:10.4018/978-1-4666-9548-1.ch022

Ruffin, T. R., Hawkins, J. M., & Lee, D. I. (2016). Organizational Leadership and Health Care Reform. In A. Normore, L. Long, & M. Javidi (Eds.), *Handbook of Research on Effective Communication, Leadership, and Conflict Resolution* (pp. 42–62). Hershey, PA: IGI Global. doi:10.4018/978-1-4666-9970-0.ch003

Sercovich, F. C. (2016). Wither Convergence?: Catching-up in an Era of Diminished Expectations. In M. Erdoğdu & B. Christiansen (Eds.), *Handbook of Research on Comparative Economic Development Perspectives on Europe and the MENA Region* (pp. 1–24). Hershey, PA: IGI Global. doi:10.4018/978-1-4666-9548-1.ch001

Singh, A., Sengupta, S., & Sharma, S. (2016). Empathy and Mindfulness: Potential Antecedents to Authentic Leadership. *International Journal of Human Capital and Information Technology Professionals*, *7*(4), 1–14. doi:10.4018/IJHCITP.2016100101

Smith, P., & Cames, O. (2016). CAMES: An Approach to Project Management Based on Action Research and the Ideal Speech Situation. In A. Normore, L. Long, & M. Javidi (Eds.), *Handbook of Research on Effective Communication, Leadership, and Conflict Resolution* (pp. 531–544). Hershey, PA: IGI Global. doi:10.4018/978-1-4666-9970-0.ch028

Stock, W. G., Barth, J., & Gremm, J. (2019). The Era After Oil: Knowledge-Intensive Cities on the Arabian Gulf. In M. Lytras, L. Daniela, & A. Visvizi (Eds.), *Knowledge-Intensive Economies and Opportunities for Social, Organizational, and Technological Growth* (pp. 63–88). Hershey, PA: IGI Global. doi:10.4018/978-1-5225-7347-0.ch005

Tayşir, N. K. (2019). Value Creation Through Social Change: An Ashoka Fellow Case. In N. Iyigun (Ed.), *Creating Business Value and Competitive Advantage With Social Entrepreneurship* (pp. 333–351). Hershey, PA: IGI Global. doi:10.4018/978-1-5225-5687-9.ch016

Thoburn, J. T. (2016). Trade and Industrial Policy for Development. In M. Erdoğdu & B. Christiansen (Eds.), *Handbook of Research on Comparative Economic Development Perspectives on Europe and the MENA Region* (pp. 25–35). Hershey, PA: IGI Global. doi:10.4018/978-1-4666-9548-1.ch002

Tran, B. (2016). Communication: The Role of the Johari Window on Effective Leadership Communication in Multinational Corporations. In A. Normore, L. Long, & M. Javidi (Eds.), *Handbook of Research on Effective Communication, Leadership, and Conflict Resolution* (pp. 405–429). Hershey, PA: IGI Global. doi:10.4018/978-1-4666-9970-0.ch022

Tuncel, C. O., & Polat, A. (2016). Nanotechnology, Long Waves, and Future of Manufacturing Industry: Comparative Analysis of European Union, East Asian Newly Industrialized Countries, and MENA Region. In M. Erdoğdu & B. Christiansen (Eds.), *Handbook of Research on Comparative Economic Development Perspectives on Europe and the MENA Region* (pp. 351–377). Hershey, PA: IGI Global. doi:10.4018/978-1-4666-9548-1.ch015

Uğurlu, M. (2016). Firm-Level Determinants of Foreign Investment and M&A Activity: Evidence from Turkey. In M. Erdoğdu & B. Christiansen (Eds.), *Handbook of Research on Comparative Economic Development Perspectives on Europe and the MENA Region* (pp. 265–292). Hershey, PA: IGI Global. doi:10.4018/978-1-4666-9548-1.ch013

Ülgen, B. (2019). The Social Entrepreneur: Personality Traits and Motivation Factors in Social Entrepreneurship. In N. Iyigun (Ed.), *Creating Business Value and Competitive Advantage With Social Entrepreneurship* (pp. 48–68). Hershey, PA: IGI Global. doi:10.4018/978-1-5225-5687-9.ch003

Vance, C. M., & Bergin, R. (2019). The Social Expat-Preneur: Examining a Growing International Career Model Supporting Global Social Entrepreneurship. In N. Iyigun (Ed.), *Creating Business Value and Competitive Advantage With Social Entrepreneurship* (pp. 186–204). Hershey, PA: IGI Global. doi:10.4018/978-1-5225-5687-9.ch009

Vatharkar, S., Gao, P., & Fomin, V. (2018). Factors Affecting Business and Information Technology Alignment at the Lower Levels of a Public Organisation. *International Journal of Healthcare Information Systems and Informatics, 13*(3), 35–48. doi:10.4018/IJHISI.2018070103

Vecchi, A., & Ricci, F. (2016). The Internationalization of the Space Industry: Anecdotal Evidence from an Italian SME. In M. Erdoğdu & B. Christiansen (Eds.), *Handbook of Research on Comparative Economic Development Perspectives on Europe and the MENA Region* (pp. 378–405). Hershey, PA: IGI Global. doi:10.4018/978-1-4666-9548-1.ch016

Venkatesh, R., Mathew, L., & Singhal, T. K. (2019). Imperatives of Business Models and Digital Transformation for Digital Services Providers. *International Journal of Business Data Communications and Networking, 15*(1), 105–124. doi:10.4018/IJBDCN.2019010107

Yalcintas, M. (2019). Creating Competitive Advantage With Social Entrepreneurship. In N. Iyigun (Ed.), *Creating Business Value and Competitive Advantage With Social Entrepreneurship* (pp. 140–160). Hershey, PA: IGI Global. doi:10.4018/978-1-5225-5687-9.ch007

Yaseen, S. G., Al-Janaydab, S., & Alc, N. A. (2018). Leadership Styles, Absorptive Capacity and Firm's Innovation. *International Journal of Knowledge Management, 14*(3), 82–100. doi:10.4018/IJKM.2018070106

Yıldız, M. L. (2019). Social Entrepreneurial Leadership for a Better World. In N. Iyigun (Ed.), *Creating Business Value and Competitive Advantage With Social Entrepreneurship* (pp. 271–293). Hershey, PA: IGI Global. doi:10.4018/978-1-5225-5687-9.ch013

Yilmaz, A., & Yilmaz, H. U. (2019). Social Value and Sociological Perspective on Social Entrepreneurship. In N. Iyigun (Ed.), *Creating Business Value and Competitive Advantage With Social Entrepreneurship* (pp. 21–47). Hershey, PA: IGI Global. doi:10.4018/978-1-5225-5687-9.ch002

Zinger, D. (2016). Developing Instructional Leadership and Communication Skills through Online Professional Development: Focusing on Rural and Urban Principals. In A. Normore, L. Long, & M. Javidi (Eds.), *Handbook of Research on Effective Communication, Leadership, and Conflict Resolution* (pp. 354–370). Hershey, PA: IGI Global. doi:10.4018/978-1-4666-9970-0.ch019

About the Contributors

Anna Visvizi is Associate Professor at Deree – The American College of Greece and Visiting Scholar at Effat University, Jeddah. A political scientist and economist, editor and research consultant, with an extensive experience in academia and the think-tank sector, Dr. Visvizi presented her work in many forums across Europe and the US, including Capitol Hill. A practiced team-worker, researcher, analyst and lecturer, Dr. Visvizi's expertise spans the following interrelated areas of research: political economy; migration; smart cities; innovation, competitiveness and policy-making, with the geographical focus of her research defined by the EU, China, and the Arab World.

Saad Bakry is Full Professor of Information Networks at King Saud University since 1989 PhD from Aston University, England, UK, 1980 CEng and FIET, UK Consultant to the Ministry of Higher Education, Saudi Arabia, Consultant to the Saudi Commission for Information and Communication Technology.

Miltiadis D. Lytras is an expert in advanced computer science, information management and innovation, editor, lecturer, and research consultant, with extensive experience in academia and the business sector in Europe and Asia. Dr. Lytras is Research Professor at Deree College – The American College of Greece and Distinguished Scientist at the King Abdulaziz University, Jeddah, Kingdom of Saudi Arabia. Dr. Lytras is a world-class expert in the fields of cognitive computing, information systems, technology enabled innovation, social networks, computers in human behavior, and knowledge management. In his work, Dr. Lytras seeks to bring together and exploit synergies among scholars and experts committed to enhancing the quality of education for all.

* * *

Haytham Abduljawad has over 30 years of experience in multinational organizations in the US and the Gulf Cooperation Council (GCC). He is the founder and CEO of Knowledge to Power consulting. He also serves as the processes & management systems division head in one of the leading national oil and gas companies in the Gulf Cooperation Council (GCC). He is the author of Knowledge to Power a Nation: Transformation to Knowledge-Based Economy through Innovation, as well as articles on knowledge management, a researcher, consultant and a speaker on Knowledge-based economy transformation strategies in developing nations and the challenge of multigenerational workforce. Haytham holds a bachelor engineering degree in Electrical and Biomedical Engineering from Manhattan College and an MBA from Bellevue University. He also completed his PhD in Human Capital Management from Bellevue University.

Odeh Al-Jayyousi is the head of Technology and Innovation Management at the Arabian Gulf University in Bahrain. He was the vice president for science and research at the Royal Scientific Society (RSS) in Jordan during 2011- 2013. He was the regional director for IUCN – West Asia regional office during 2004-2011. Previously, he was a university professor in water resources and environment during 1994-2004. He was an academic director for M.Sc. Programme in transformational management in UK (1999-2004). In 2014, he worked as a consultant with UN ESWCA on Technology for Sustainable Development. His research interests include Technology for Sustainable Development, appropriate technologies, science policy, technology assessment, technology oversight, water, and environmental policy, and innovation policy. He was a consultant with EU, ESCWA, World Bank, GTZ and USAID. He published several articles in international peer-reviewed journals including several books including two recent books; i.e., Islam and sustainable development (2012), UK and Renewable Energy and Knowledge Management, 2015. He was the Dean of Scientific Research at Applied Science University, Jordan. He is a founding board member in Jordan Green Building Council and Fair Trade Jordan. He is a member of the Advisory Board for MSc program in IWRM between Cologne and Jordan University; Board member of several think tanks WANA Forum, and Social Forum and Economic Forum. He was a keynote speaker in many global events including Global Thinkers Forum and Sustainability Forum. He is a recipient of the award for scientists and social innovations, 2014.

Yousef Alabbasi is a Ph.D. Student at University of New England.

Ahmed El-Kholei is Professor of Urban Planning, Department of Natural Resources and Environment, AGU. He holds a Ph.D. degree in Regional Planning from the University of Illinois at Urbana-Champaign, USA; MSc degree in Urban Planning and Policy Development from Rutgers; in addition to MA and BA degrees in Architecture from Faculty of Fine Arts, Cairo, Egypt. He taught at the University of Illinois at Urbana-Champaign and Michigan State University, USA, and Arab Academy for Science and technology, Menofia University, and Ain Shams University, Egypt. He provided technical assistance in participatory strategic and action planning for urban and regional planning, and national environmental management programmes. He participated in designing several national and regional development projects. He led the evaluation of development and environmental projects. He worked with several agencies including, but not limited to, UNDP, UNEP, FAO, IFAD, WFP, GEF, ECA, AfDB, LAS, and GIZ. In 2006, the UN Secretary-General appointed him as a member of the Advisory Group for Central Emergency Relief Fund (CERF) of OCHA for two years for his outstanding expertise in the fields of environmental management and urban planning. He coordinated the activities to prepare Egypt's National Environmental Action Plan. His research interests include urban and regional planning theories, methods and models, environmental management, disaster risk reduction, and sustainable development.

Saeed-Ul Hassan is the Director of Scientometrics Lab and a faculty member at Information Technology University (ITU), a former Post-Doctorate Fellow at the United Nations University – with more than 15 years of hands-on experience of advance statistical techniques, artificial intelligence and software development client work. He earned his PhD in the field Information Management from Asian Institute of Technology, Thailand. He has also served as a Research Fellow at National Institute of Informatics in Japan. Dr. Saeed's research interests lie within the areas of Data Science, Scientometrics, Bibliometric Tools for Evidence-based Research Policy Formulation, Information Retrieval and Text Mining. He has published more than 30 papers in reputed international journals and conference proceedings including highly reputed venues like WWW, JCDL, Journal of Infometrics and Scientometrics. Dr. Saeed is also the recipient of James A. Linen III Memorial Award in recognition of his outstanding academic performance. More recently, he has been awarded Eugene Garfield Honorable Mention Award for Innovation in Citation Analysis by Clarivate Analytics, Thomson Reuters.

Jung-Tsung Ho (Ph.D., Business Administration, National Central University in Taiwan) is currently focusing his studies on the subject of innovation and entrepreneurship by the approach of text analytics and social network analysis.

Fahad Sabah has done BS Electrical Engineering (2012) form Punjab University, MS in Computer Sciences(2014) from Information Technology University, Punjab, Pakistan. He is a full-time Faculty Member at Superior University Lahore since November, 2014. His Current Research Interests are Data Mining and Scientometrics. Currently, working as an Advisor IEEE-Academic Student Branch in Superior University Lahore (IEEE-SUL). Had also acted as manager of final year project and Program In-Charge of CS&IT at Superior University, Lahore.

Kamaljeet Sandhu's teaching and research expertise are in Accounting, Information Systems & IT Adoption, Management Accounting and Corporate Governance Adoption, and E-Learning Acceptance at Universities. Dr. Sandhu is an active, experienced researcher & passionate about research in Digital Accounting Innovation & Strategy, Blockchain, Business Analytics, AI, IoT, Cryptocurrencies, IT Startups, FinTech, Cybersecurity, Corporate Social Responsibility, Supply Chain, ERP, Digital Health, SMEs, & Entrepreneurship.

Chien-wen Shen is an Associate Professor in the Department of Business Administration at National Central University (NCU). He also serves as the Director of the Yunus Social Business Center, Director of the Center for Media and Social Impact, and Division Director of the University Social Responsibility Office at NCU. He received his M.S. and Ph.D. in Industrial Engineering and Management Sciences from Northwestern University. His research interests include business analytics, information management, decision science, and social value. Dr. Shen has published over 90 papers and presented numerous lectures/speeches/workshops. He has also been participated with more than 30 projects supported by Ministry of Science and Technology, Ministry of Transportation and Communications, Ministry of Education, and National Science Foundation. He is also an Executive Director of the Social Impact Institute of Taiwan and a director of the Chinese Enterprise Resource Planning Society.

Saira Hanif Soroya is serving as Assistant Professor at the Department of Information Management, University of the Punjab, Lahore. She has published her research work in well reputed International and National journals. In addition, she has presented research papers in International and national conferences as well. She is a professional trainer and served as a resource person in several trainings in various universities of Pakistan. She has also served as managing editor of the "International Journal of Information & Management Sciences". She is the first female "Brand Ambassador" of Emerald Publishing in Pakistan.

Yaşar Serhat Yaşgül is an associate professor of Economics at Marmara University. He studied at the Marmara University (PhD in International Economics and Master in Econometrics), and his research has centered on international political economy, in particular intellectual property rights. More recently, he has moved into studying of development issues, particularly in the BRIC countries and Turkey.

Index

S

T

U

W

www.ingramcontent.com/pod-product-compliance
Lightning Source LLC
LaVergne TN
LVHW081321300926
849161LV00008B/1839